Road Atlas
for the
Annular
Solar Eclipse
of
2023

Black and White Edition

Fred Espenak

Edition 1.0
July 2018

Road Atlas for the Annular Solar Eclipse of 2023 – Black and White Edition

Astropixels Publishing
P.O. Box 16197
Portal, AZ 85632

Astropixels Publishing Website: *astropixels.com/pubs*

This book may be ordered at: *astropixels.com/pubs/Atlas2023.html*

Printed in the United States of America

ISBN 978-1-941983-16-4

Astropixels Publication: AP016 (Version 1.0b)

First Edition

Front Cover: A map of the 2023 eclipse path in the vicinity of the Texas Gulf Coast illustrates the cartographic features found in the ***Road Atlas for the Annular Solar Eclipse of 2023 – Black and White Edition***. Map copyright © 2018 by Fred Espenak. More about the eclipse can be found at:

eclipsewise.com/solar/SEnews/ASE2023.html

Back Cover Photo of Fred Espenak: Copyright © 2016 by Fred Espenak

Table of Contents

The last annular solar eclipse visible from the contiguous USA was on May 20, 2012.
The image above was shot from Elida, NM, where the annular phase occurred about 10 minutes before sunset.
© 2012 F. Espenak, www.MrEclipse.com

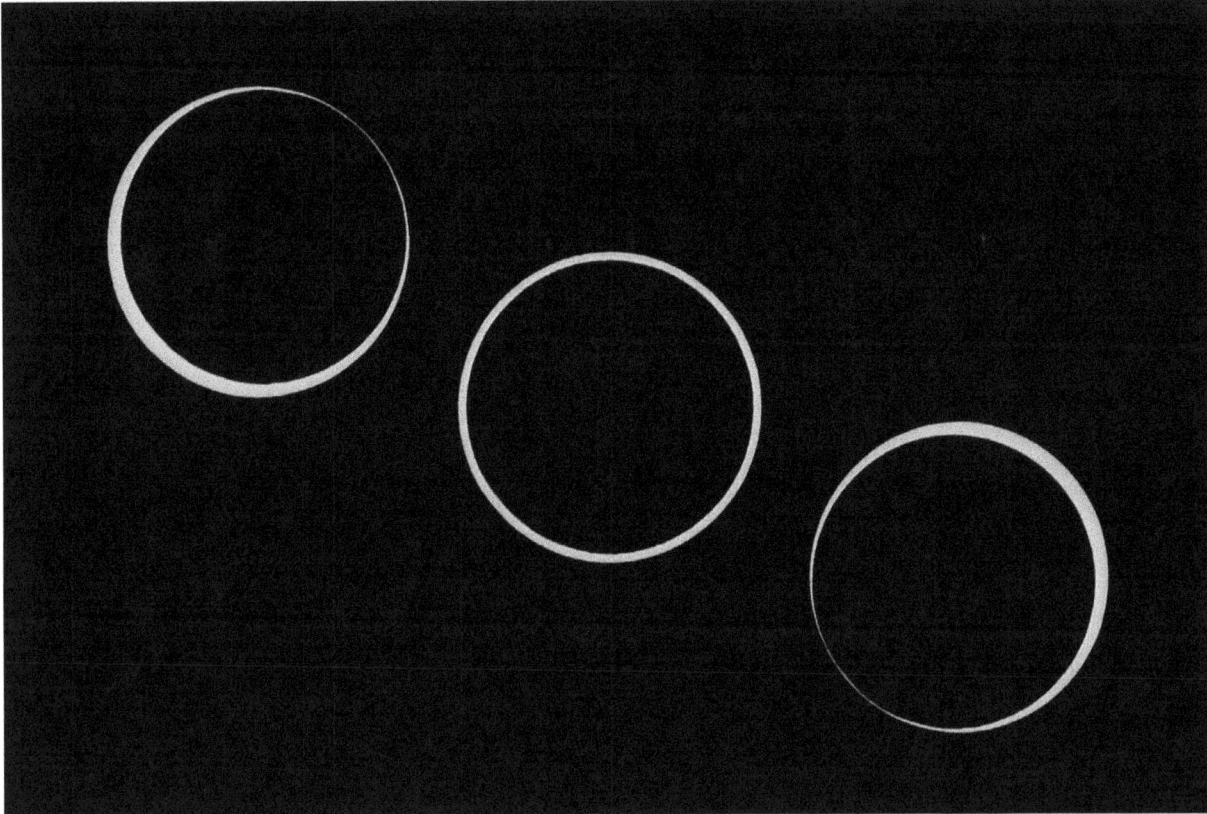

The rapidly changing geometry of the Sun and Moon over five minutes is revealed in the above composite of annularity.
Left: just as annularity begins. Middle: mid-annularity. Right: just as annularity ends.
This sequence was captured during the annular solar eclipse of October 3, 2005 from Carrascosa del Campo, Spain.
© 2005 F. Espenak, www.MrEclipse.com

1 – Introduction

On Saturday, October 14, 2023, an annular eclipse of the Sun will be visible from the United States for the first time since May 20, 2012. Although a partial eclipse will be seen from all of North America, the annular phase in which the Moon's disk is completely silhouetted by the Sun (i.e., *annularity*) is only visible from a narrow path of the Moon's antumbral shadow as it sweeps through western USA, Mexico, Central and South America.

Because a bright ring of sunlight will surround the Moon during *annularity*, the Sun's faint solar corona will remain hidden. The only safe way to look directly at the eclipse is through special solar filters, such as "eclipse glasses" or hand-held solar viewers. This includes the annular phase of the eclipse.

The course of the Moon's shadow begins in the North Pacific and crosses the western USA, traversing parts of nine states[1]: Oregon, California, Idaho, Nevada, Utah, Colorado, Arizona, New Mexico, and Texas. The central path also crosses Mexico's Yucatan Peninsula, and parts of six Central American nations: Guatemala, Belize, Honduras, Nicaragua, Costa Rica, and Panama. Finally, the path of annularity sweeps across Colombia and Brazil before ending in the Atlantic Ocean. The width of the *path of annularity* ranges from 137 miles (220 km) in Oregon to 115 miles (184 km) in Mexico and Belize.

The duration of annularity also varies along the path being longest in Panama (5 minutes 17 seconds) and shortest in east Brazil (4 minutes 20 seconds). These durations are for the middle of the path (*central line)*. As one moves away from the *central line*, the duration of *annularity* decreases. It happens slowly at first but drops rapidly to *zero* at the northern or southern edges or limits of the *path of annularity*.

Observing from the central line of the eclipse offers the longest possible duration of annularity. Alternatively, positioning an observer just inside the northern or southern limits shortens annularity but prolongs the appearance of Baily's beads. In either case, this collection of maps will assist eclipse observers in locating the exact position of the path of annularity.

---

[1] Only a small corner of southwestern Idaho lies in the path of annularity.

2 – Local Circumstances for the Eclipse

Local circumstances for the eclipse appear in Tables 1 through 6. All times are given in local time and include Daylight Saving Time (where applicable). The Sun's altitude is given at the instant of maximum eclipse. *Eclipse Magnitude* is the fraction (in percent) of the Sun's *diameter* covered by the Moon at maximum eclipse

The circumstances for a number of U.S. cities appear in Table 1 (annular eclipse) and Table 2 (partial eclipse). Cities in Canada are found in Table 3 (partial eclipse), and Mexico and Central America in Table 4 (partial eclipse). Finally, circumstances for major cities in South America are listed in Tables 5 (partial eclipse) and 6 (annular eclipse).

Table 1 - Local Circumstances for the Annular Eclipse in the USA

State/City	Partial Eclipse Begins	Annular Eclipse Begins	Annular Eclipse Ends	Partial Eclipse Ends	Duration Of Annularity	Sun Altitude
OREGON						
Coos Bay	08:04:57 am	09:16:01 am	09:20:03 am	10:38:39 am	04m03s	18°
Corvallis	08:05:35 am	09:17:57 am	09:19:44 am	10:39:26 am	01m47s	18°
Eugene	08:05:27 am	09:16:59 am	09:20:47 am	10:39:46 am	03m49s	18°
Klamath Falls	08:05:17 am	09:17:57 am	09:21:18 am	10:41:47 am	03m21s	20°
Medford	08:05:01 am	09:18:21 am	09:19:20 am	10:40:26 am	00m59s	19°
Roseburg	08:05:07 am	09:16:25 am	09:20:47 am	10:39:41 am	04m21s	18°
Springfield	08:05:27 am	09:17:00 am	09:20:51 am	10:39:50 am	03m51s	18°
NEVADA						
Elko	08:07:14 am	09:22:27 am	09:26:46 am	10:50:07 am	04m19s	25°
NEW MEXICO						
Albuquerque	09:13:08 am	10:34:26 am	10:39:16 am	12:09:21 pm	04m50s	36°
Artesia	09:15:58 am	10:39:56 am	10:43:33 am	12:16:23 pm	03m36s	40°
Carlsbad	09:16:21 am	10:41:33 am	10:43:12 am	12:17:19 pm	01m39s	40°
Farmington	09:11:40 am	10:31:49 am	10:36:08 am	12:04:55 pm	04m19s	34°
Gallup	09:11:38 am	10:33:02 am	10:35:07 am	12:05:22 pm	02m05s	34°
Hobbs	09:17:07 am	10:41:10 am	10:46:05 am	12:18:53 pm	04m55s	41°
Los Alamos	09:13:14 am	10:35:13 am	10:38:22 am	12:08:58 pm	03m09s	36°
Roswell	09:15:35 am	10:38:43 am	10:43:24 am	12:15:22 pm	04m41s	39°
Santa Fe	09:13:25 am	10:35:51 am	10:38:39 am	12:09:43 pm	02m48s	36°
TEXAS						
Alamo	10:23:43 am	11:51:51 am	11:56:25 am	01:32:50 pm	04m34s	47°
Alice	10:25:45 am	11:55:09 am	11:59:18 am	01:36:57 pm	04m09s	49°
Andrews	10:17:50 am	11:42:22 am	11:47:18 am	01:20:31 pm	04m55s	42°
Beeville	10:25:30 am	11:54:23 am	11:59:18 am	01:36:21 pm	04m55s	49°
Big Spring	10:18:52 am	11:44:42 am	11:48:11 am	01:22:34 pm	03m28s	42°
Corpus Christi	10:26:27 am	11:55:46 am	12:00:48 pm	01:38:13 pm	05m02s	49°
Ingleside	10:26:34 am	11:55:59 am	12:00:55 pm	01:38:23 pm	04m56s	49°
Kerrville	10:22:38 am	11:50:20 am	11:54:35 am	01:30:38 pm	04m15s	46°
Kingsville	10:26:13 am	11:55:52 am	11:59:57 am	01:37:49 pm	04m04s	49°
Lamesa	10:18:08 am	11:43:42 am	11:46:47 am	01:20:54 pm	03m05s	42°
Midland	10:18:27 am	11:43:22 am	11:48:17 am	01:21:52 pm	04m55s	42°
Odessa	10:18:15 am	11:43:07 am	11:47:57 am	01:21:32 pm	04m49s	42°
Portland	10:26:26 am	11:55:45 am	12:00:45 pm	01:38:09 pm	05m00s	49°
Robstown	10:26:10 am	11:55:24 am	12:00:18 pm	01:37:42 pm	04m54s	49°
San Angelo	10:20:18 am	11:47:11 am	11:50:21 am	01:25:43 pm	03m10s	44°
San Antonio	10:23:48 am	11:52:03 am	11:56:28 am	01:32:59 pm	04m25s	47°
Schertz	10:23:58 am	11:52:51 am	11:56:07 am	01:33:13 pm	03m15s	47°
Uvalde	10:22:40 am	11:50:34 am	11:54:29 am	12:30:55 pm	03m55s	46°

Table 2 - Local Circumstances for the Partial Eclipse in the USA

City, State	Partial Eclipse Begins	Maximum Eclipse	Partial Eclipse Ends	Eclipse Magnitude	Sun Altitude
Albany, NY	12:10:12 pm	01:20:35 pm	02:31:16 pm	32%	38°
Atlanta, GA	11:43:10 am	01:12:18 pm	02:45:09 pm	62%	48°
Augusta, ME	12:22:41 pm	01:25:52 pm	02:28:39 pm	25%	36°
Austin, TX	10:23:59 am	11:54:22 am	01:32:48 pm	93%	47°
Baltimore, MD	12:01:26 pm	01:19:25 pm	02:38:22 pm	41%	42°
Baton Rouge, LA	10:32:31 am	12:04:56 pm	01:43:35 pm	80%	50°
Birmingham, AL	10:38:29 am	12:08:16 pm	01:42:40 pm	67%	48°
Bismarck, ND	10:21:23 am	11:38:39 am	01:01:25 pm	64%	30°
Boise, ID	09:07:43 am	10:23:59 am	11:47:49 am	90%	23°
Boston, MA	12:18:07 pm	01:25:57 pm	02:33:29 pm	29%	38°
Charleston, SC	11:53:29 am	01:22:09 pm	02:52:52 pm	56%	49°
Chicago, IL	10:37:21 am	11:58:06 am	01:22:45 pm	54%	39°
Columbus, OH	11:46:05 am	01:07:18 pm	02:31:22 pm	50%	42°
Concord, NH	12:16:54 pm	01:23:59 pm	02:30:54 pm	28%	37°
Dallas, TX	10:23:35 am	11:52:50 am	01:29:39 pm	86%	45°
Denver, CO	09:13:59 am	10:36:07 am	12:05:59 pm	85%	33°
Des Moines, IA	10:27:29 am	11:49:38 am	01:17:11 pm	64%	38°
Detroit, MI	11:46:45 am	01:04:55 pm	02:25:43 pm	46%	39°
El Paso, TX	09:15:05 am	10:39:57 am	12:14:01 pm	90%	39°
Fargo, ND	10:26:13 am	11:43:11 am	01:05:00 pm	58%	31°
Hartford, CT	12:13:02 pm	01:23:37 pm	02:34:14 pm	32%	39°
Houston, TX	10:27:10 am	11:58:53 am	01:38:05 pm	90%	49°
Jackson, MS	10:33:03 am	12:04:04 pm	01:40:54 pm	75%	48°
Kansas City, MO	10:25:27 am	11:50:04 am	01:20:47 pm	70%	40°
Knoxville, TN	11:43:45 am	01:10:14 pm	02:40:10 pm	58%	46°
Las Vegas, NV	08:08:02 am	09:26:38 am	10:54:10 am	87%	29°
Lincoln, NE	10:23:02 am	11:45:59 am	01:15:07 pm	70%	37°
Little Rock, AR	10:28:57 am	11:57:38 am	01:32:31 pm	74%	45°
Los Angeles, CA	08:07:57 am	09:24:31 am	10:50:07 am	78%	28°
Louisville, KY	11:40:02 am	01:04:26 pm	02:32:41 pm	57%	43°
Madison, WI	10:34:48 am	11:54:37 am	01:18:36 pm	55%	38°
Memphis, TN	10:32:24 am	12:00:48 pm	01:34:42 pm	70%	45°
Miami, FL	11:57:20 am	01:33:49 pm	03:11:50 pm	67%	55°
Milwaukee, WI	10:37:26 am	11:56:50 am	01:20:02 pm	53%	38°
Nashville, TN	10:38:01 am	12:05:02 pm	01:36:26 pm	62%	45°
New Orleans, LA	10:34:32 am	12:07:32 pm	01:46:27 pm	79%	51°
New York, NY	12:08:53 pm	01:22:27 pm	02:36:18 pm	35%	40°
Oklahoma City, OK	10:21:48 am	11:48:58 am	01:23:29 pm	82%	42°
Philadelphia, PA	12:05:31 pm	01:21:17 pm	02:37:38 pm	37%	41°
Phoenix, AZ	08:10:43 am	09:31:34 am	11:01:40 am	85%	33°
Portland, ME	12:20:53 pm	01:25:43 pm	02:30:11 pm	26%	36°
Portland, OR	08:06:10 am	09:19:25 am	10:39:50 am	91%	18°
Providence, RI	12:16:55 pm	01:25:58 pm	02:34:47 pm	30%	38°
Raleigh, NC	11:56:00 am	01:20:14 pm	02:46:04 pm	49%	46°
Richmond, VA	11:59:01 am	01:20:03 pm	02:42:20 pm	44%	44°
Salem, OR	08:05:48 am	09:19:04 am	10:39:36 am	93%	18°
Salt Lake City, UT	09:09:12 am	10:28:17 am	11:55:24 am	92%	28°
San Francisco, CA	08:05:17 am	09:19:29 am	10:42:07 am	83%	22°
Seattle, WA	08:07:31 am	09:20:11 am	10:39:35 am	86%	17°
Springfield, IL	10:33:11 am	11:56:46 am	01:25:05 pm	61%	41°
Washington, DC	12:00:13 pm	01:19:06 pm	02:39:06 pm	42%	43°

Table 3 - Local Circumstances for the Partial Eclipse in Canada

City, State	Partial Eclipse Begins	Maximum Eclipse	Partial Eclipse Ends	Eclipse Magnitude	Sun Altitude
Calgary, AB	09:14:13 am	10:26:51 am	11:45:09 am	70%	20°
Charlottetown, PE	01:45:10 pm	02:35:23 pm	03:24:43 pm	15%	31°
Edmonton, AB	09:17:33 am	10:28:19 am	11:44:03 am	63%	18°
Fredericton, NB	01:33:13 pm	02:29:25 pm	03:24:57 pm	19%	33°
Halifax, NS	01:43:13 pm	02:36:48 pm	03:29:21 pm	17%	33°
Hamilton, ON	11:54:26 am	01:09:20 pm	02:25:57 pm	40%	39°
Inuvik, NW	-	10:33:02 am	11:32:22 am	44%	3°
Iqaluit, NU	12:37:38 pm	01:11:39 pm	01:45:38 pm	9%	17°
Lethbridge, AB	09:13:38 am	10:27:25 am	11:47:09 am	72%	21°
London, ON	11:51:04 am	01:07:17 pm	02:25:36 pm	43%	39°
Montréal, QC	12:11:56 pm	01:17:46 pm	02:23:53 pm	28%	36°
Ottawa, ON	12:06:02 pm	01:14:14 pm	02:23:09 pm	31%	36°
Prince George, BC	08:13:26 am	09:22:50 am	10:37:27 am	71%	13°
Québec, QC	12:19:18 pm	01:20:25 pm	02:21:27 pm	24%	34°
Regina, SK	10:20:47 am	11:34:37 am	12:53:32 pm	61%	25°
Saint John's, NF	02:54:11 pm	03:21:28 pm	03:48:20 pm	5%	25°
Saskatoon, SK	10:20:55 am	11:33:10 am	12:50:14 pm	60%	23°
Sault St. Marie, ON	11:46:30 am	12:59:48 pm	02:15:34 pm	42%	35°
Thunder Bay, ON	11:38:51 am	12:52:06 pm	02:08:31 pm	46%	32°
Toronto, ON	11:55:45 am	01:09:44 pm	02:25:19 pm	39%	38°
Vancouver, BC	08:08:32 am	09:20:18 am	10:38:25 am	82%	15°
Victoria, BC	11:07:47 am	12:19:50 pm	01:38:25 pm	85%	16°
Winnipeg, MB	10:28:33 am	11:42:16 am	01:00:14 pm	53%	29°
Windsor, ON	11:46:48 am	01:04:59 pm	02:25:49 pm	46%	39°
Yellowknife, NW	09:31:21 am	10:33:36 am	11:38:44 am	46%	12°
Whitehorse, YU	-	09:23:49 am	10:31:00 am	62%	4°

Table 4 - Local Circumstances for the Partial Eclipse in Mexico and Central America

City	Partial Eclipse Begins	Maximum Eclipse	Partial Eclipse Ends	Eclipse Magnitude	Sun Altitude
MEXICO					
Ciudad Juarez	09:15:07 am	10:39:59 am	12:14:03 pm	90%	39°
Guadalajara	09:31:04 am	10:59:43 am	12:37:16 pm	71%	52°
Leon	09:31:30 am	11:01:51 am	12:40:56 pm	75%	53°
Mexico City	09:36:26 am	11:09:22 am	12:50:22 pm	77%	57°
Monterrey	09:25:50 am	10:56:38 am	12:36:09 pm	87%	50°
Puebla	09:38:02 am	11:11:52 am	12:53:29 pm	78%	58°
San Luis Potosi	09:30:18 am	11:01:14 am	12:40:52 pm	79%	53°
Tijuana	09:09:21 am	10:26:37 am	11:53:06 am	76%	30°
Toluca	09:36:22 am	11:08:45 am	12:49:18 pm	76%	57°
Torreon	09:23:28 am	10:51:50 am	12:29:24 pm	81%	47°
COSTA RICA					
San Jose	10:15:11 am	12:00:51 pm	01:46:06 pm	93%	69°
EL SALVADOR					
San Salvador	09:58:52 am	11:40:56 am	01:26:32 pm	88%	68°
GUATEMALA					
Guatemala	09:55:05 am	11:36:04 am	01:21:26 pm	87%	67°
NICARAGUA					
Managua	10:06:36 am	11:50:44 am	01:36:27 pm	92%	69°
PANAMA					
Panama City	11:26:05 am	01:13:20 pm	02:56:56 pm	94%	66°

Table 5 - Local Circumstances for the Partial Eclipse in South America

Country/City	Partial Eclipse Begins	Maximum Eclipse	Partial Eclipse Ends	Eclipse Magnitude	Sun Altitude
ARGENTINA					
Buenos Aires	04:02:54 pm	04:43:42 pm	05:22:23 pm	13%	29°
BOLIVIA					
La Paz	01:55:42 pm	03:25:01 pm	04:44:01 pm	60%	44°
BRAZIL					
Belém	03:04:24 pm	04:32:44 pm	05:47:23 pm	83%	22°
Belo Horizonte	03:37:12 pm	04:49:32 pm	05:53:14 pm	59%	15°
Brasília	03:25:39 pm	04:45:41 pm	05:55:01 pm	72%	19°
Fortaleza	03:23:34 pm	04:42:44 pm	-	89%	10°
Manaus	02:40:35 pm	04:19:29 pm	05:43:39 pm	93%	37°
Recife	03:31:55 pm	04:47:17 pm	-	92%	6°
Rio de Janeiro	03:42:59 pm	04:50:31 pm	05:50:42 pm	51%	14°
Santos	03:41:53 pm	04:49:27 pm	05:49:45 pm	48%	17°
São Paulo	03:40:50 pm	04:49:17 pm	05:50:16 pm	49%	18°
CHILE					
Santiago	03:56:12 pm	04:33:57 pm	05:10:12 pm	10%	41°
COLOMBIA					
Bogota	11:48:20 am	01:36:15 pm	03:15:26 pm	93%	59°
ECUADOR					
Quito	11:51:15 am	01:37:42 pm	03:16:57 pm	85%	64°
PARAGUAY					
Asuncion	03:32:52 pm	04:42:41 pm	05:45:20 pm	41%	29°
PERU					
Lima	12:29:13 pm	02:04:21 pm	03:31:02 pm	60%	58°
URUGUAY					
Montevideo	04:05:00 pm	04:45:02 pm	05:22:59 pm	13%	26°
VENEZUELA					
Caracas	12:56:00 pm	02:39:09 pm	04:11:30 pm	70%	49°

Table 6 - Local Circumstances for the Annular Eclipse in Central & South America

Country/City	Partial Eclipse Begins	Annular Eclipse Begins	Annular Eclipse Ends	Partial Eclipse Ends	Duration Annularity	Sun Altitude
MEXICO						
Campeche	10:45:24 am	12:22:22 pm	12:26:57 pm	02:09:23 pm	04m36s	61°
Chetumal	10:50:59 am	12:29:44 pm	12:34:08 pm	02:17:09 pm	04m24s	63°
BELIZE						
Belize City	09:52:53 am	11:31:44 am	11:36:55 am	01:19:46 pm	05m11s	64°
HONDURAS						
La Ceiba	09:58:14 am	11:38:24 am	11:43:36 am	01:26:42 pm	05m12s	66°
NICARAGUA						
Bluefields	10:11:12 am	11:53:55 am	11:59:10 am	01:41:58 pm	05m15s	68°
COLOMBIA						
Buenaventura	11:43:22 am	01:29:31 pm	01:32:53 pm	03:12:01 pm	03m23s	63°
Cali	11:45:37 am	01:31:35 pm	01:35:18 pm	03:13:46 pm	03m43s	62°
Cartago	11:43:46 am	01:30:13 pm	01:33:15 pm	03:12:11 pm	03m02s	61°
Neiva	11:49:33 am	01:34:48 pm	01:39:54 pm	03:16:42 pm	05m06s	60°
Palmira	11:45:57 am	01:31:27 pm	01:36:09 pm	03:14:01 pm	04m42s	62°
BRAZIL						
Campina Grande	03:30:04 pm	04:44:55 pm	04:47:59 pm	-	03m04s	7°
Crato	03:25:56 pm	04:43:09 pm	04:47:00 pm	-	03m52s	11°
Joao Pessoa	03:31:03 pm	04:45:09 pm	04:48:14 pm	-	03m05s	6°
Natal	03:29:26 pm	04:43:54 pm	04:47:27 pm	-	03m33s	6°
Patos	03:28:18 pm	04:43:50 pm	04:47:49 pm	-	03m59s	8°
Santa Rita	03:30:58 pm	04:45:09 pm	04:48:12 pm	-	03m03s	6°

9

The *2023 Annular Solar Eclipse Circumstances Calculator* is an interactive web page that can quickly calculate the local circumstances for the eclipse from any geographic location not included in Tables 1 through 6. The *Calculator* is located at: *www.EclipseWise.com/solar/SEcirc/2001-2100/SE2023Oct14Acirc.html*

3 – Global Map of the Eclipse

The global map of Earth (page 12) illustrates the geographic extent of solar eclipse visibility. The limits of the Moon's penumbral shadow indicate the region where the partial solar eclipse is visible. This irregular, saddle-shaped region covers much of the daylight hemisphere of Earth and consists of several distinct zones. Great loops at the eastern and western ends (magenta curves) identify the areas where the eclipse begins/ends at sunrise and sunset, respectively. Bisecting the 'eclipse begins/ends at sunrise and sunset' loops is the curve of maximum eclipse at sunrise (western loop) and sunset (eastern loop).

The curves of maximum eclipse are given at each half-hour Universal Time. They run between the northern and southern limits of the penumbra. The curves of constant eclipse magnitude[2] identify points where the maximum eclipse magnitude has a constant value of 0.2, 0.4, 0.6 and 0.8. The annular eclipse path appears as a solid orange band. It is only within this zone that the annular eclipse is visible.

Additional data pertinent to the eclipse appear in the corners and at the bottom of the map. For a complete description of this information, see: *www.eclipsewise.com/oh/oh-help/SEdiskkey.html*

4 – Overview Maps of the Path of Annularity

The three maps on pages 14-16 give a broad overview of the annular eclipse path through the USA, Mexico, Central and South America. The yellow lines running across the eclipse path mark the position of mid-eclipse at 10-minute intervals. The times are given in local time at each position along with the duration of annularity on the central line, and the altitude of the Sun above the horizon at that instant.

5 – Detailed Maps of the Path of Annularity

A detailed series of 29 maps (pages 17-45) covers the entire land-based path of annularity. The *Table of Contents* (page iii) can be used to quickly navigate to the map of interest since it lists the location of each map.

The map scale is approximately 1:1,800,000, which corresponds to 1 inch ≈ 28 miles (1 cm ≈ 18 km).[3] This large scale shows both major and minor roads, towns and cities, rivers, lakes, parks, national forests, wilderness areas and mountain ranges. A 50 mile (50 km) reference scale appears at the bottom of each map.

The path of annularity is depicted as a lightly shaded region with the northern and southern limits clearly labeled. The annular phase can be seen only inside this path (a partial eclipse is visible outside the path). The closer one gets to the central line, the longer the annular phase lasts. Gray lines inside the path mark the duration of the annular eclipse in 30-second steps, making it easy to estimate the duration anywhere.

The local time of mid-eclipse is marked by a series of yellow lines crossing the eclipse path every 5 minutes. Abbreviations for local times are: PDT = Pacific Daylight Time, MDT = Mountain Daylight Time, CDT = Central Daylight Time, EDT = Eastern Daylight Time, EST = Eastern Standard Time, AMT = Amazon Time, and BRT = Brazil Standard Time. Eclipse circumstances on the central line are labeled with the local time of mid-eclipse, duration of annularity (minutes and seconds) and altitude of the Sun.

All maps were produced using Google Maps as the underlying map with overlying eclipse graphics generated using Javascript code. A web page is available to the user for examining any part of the 2023 eclipse path at a range of zoom magnifications. An added benefit of the web page is that it automatically calculates the local circumstances for any point the user chooses. For more information on the interactive 2023 eclipse map:

www.eclipsewise.com/solar/SEgmap/2001-2100/SE2023Oct14Agmap.html

---

[2] Eclipse magnitude is defined as the fraction of the Sun's diameter occulted by the Moon.

[3] Because of the Mercator map projection, the actual scale on a given map can vary from this value.

Global Map of the 2023 Annular Eclipse

Annular Solar Eclipse of 2023 Oct 14

Greatest Eclipse = 18:00:40.6 TD   (= 17:59:29.3 UT1)

Eclipse Magnitude = 0.9520   Saros Series = 134
Gamma = 0.3753   Saros Member = 44 of 71

Sun at Greatest Eclipse
(Geocentric Coordinates)
R.A. = 13h18m05.4s
Dec. = -08°14'36.7"
S.D. = 00°16'02.0"
H.P. = 00°00'08.8"

Moon at Greatest Eclipse
(Geocentric Coordinates)
R.A. = 13h18m44.3s
Dec. = -07°56'18.9"
S.D. = 00°15'02.9"
H.P. = 00°55'13.8"

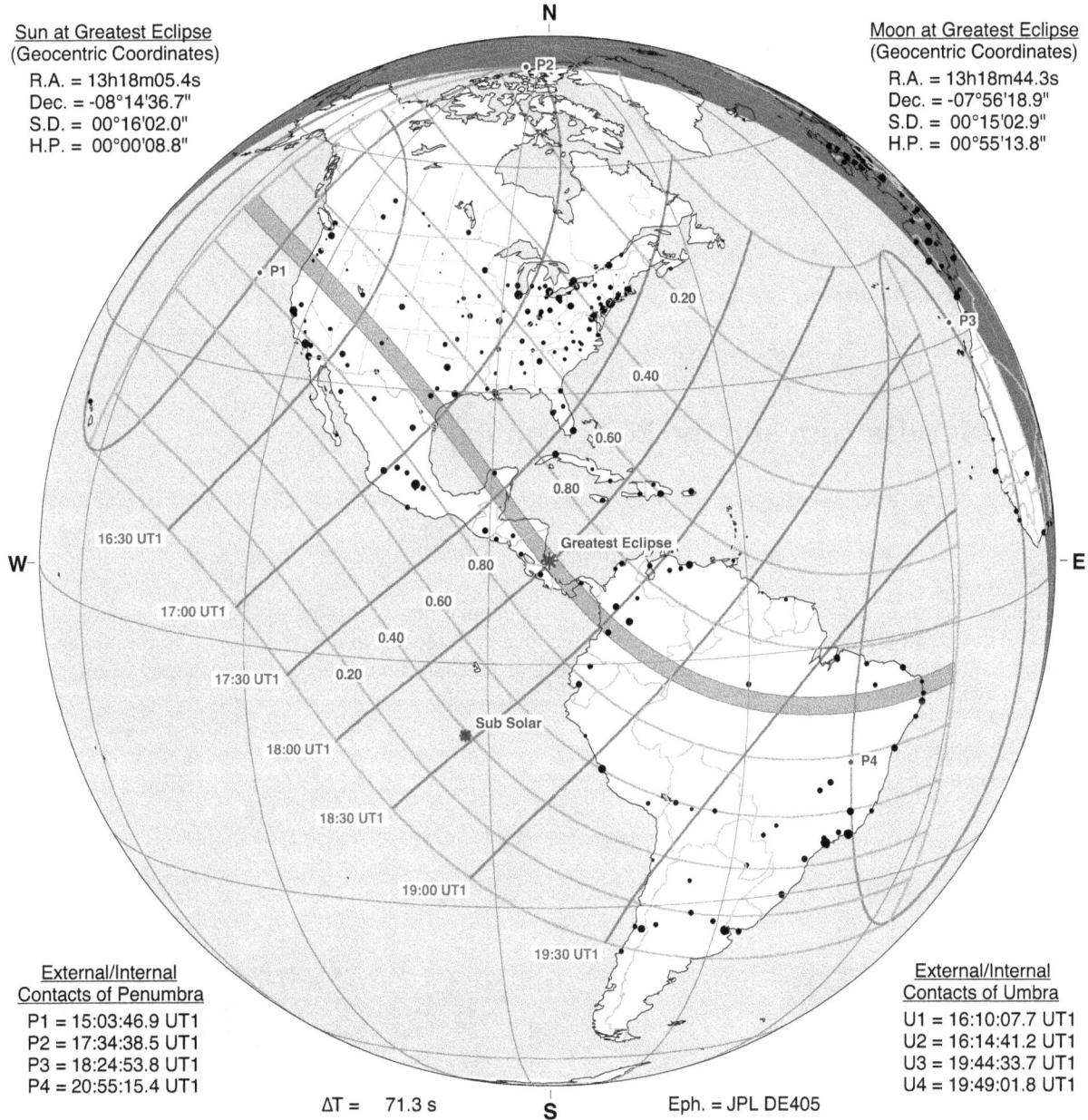

N

P2
P1
0.20
0.40
0.60
0.80
Greatest Eclipse
0.80
0.60
0.40
0.20
P3
Sub Solar
P4

16:30 UT1
17:00 UT1
17:30 UT1
18:00 UT1
18:30 UT1
19:00 UT1
19:30 UT1

W
E

External/Internal
Contacts of Penumbra
P1 = 15:03:46.9 UT1
P2 = 17:34:38.5 UT1
P3 = 18:24:53.8 UT1
P4 = 20:55:15.4 UT1

External/Internal
Contacts of Umbra
U1 = 16:10:07.7 UT1
U2 = 16:14:41.2 UT1
U3 = 19:44:33.7 UT1
U4 = 19:49:01.8 UT1

ΔT = 71.3 s   S   Eph. = JPL DE405

Circumstances at Greatest Eclipse: 17:59:29.3 UT1

Lat. = 11°22.1'N   Sun Alt. = 67.9°
Long. = 083°06.1'W   Sun Azm. = 208.0°
Path Width = 187.4 km   Duration = 05m17.2s

Circumstances at Greatest Duration: 18:13:09.2 UT1

Lat. = 08°14.6'N   Sun Alt. = 66.8°
Long. = 080°24.1'W   Sun Azm. = 225.1°
Path Width = 191.1 km   Duration = 05m17.8s

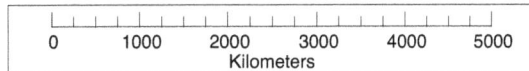

©2016 F. Espenak
www.EclipseWise.com

0	1000	2000	3000	4000	5000

Kilometers

Map courtesy of "21st Century Canon of Solar Eclipses" by Fred Espenak (2017)

11

6 – EclipseWise.com Web Site

For many years the *NASA Eclipse Web Site* was the primary Internet resource for predictions and information on eclipses of the Sun and Moon. *EclipseWise.com* has now superseded the *NASA Eclipse Web Site* with new and improved features.

EclipseWise.com has individual web pages, maps and diagrams for every solar and lunar eclipse from 2000 BCE to 3000 CE. This covers to 11,898 solar eclipses and 12,064 lunar eclipses. Much of the design, layout and graphics were inspired by the recent publications *Thousand Year Canon of Solar Eclipses 1501 to 2500* and the *Thousand Year Canon of Lunar Eclipses 1501 to 2500*. (See: www.astropixels.com/pubs)

The graphical user interface used by *EclipseWise.com* offers an intuitive way of accessing eclipse predictions. For example, the home page presents a concise preview of all upcoming solar and lunar eclipses over several years. Each small eclipse diagram gives a quick preview of an eclipse and links to a dedicated page for that particular eclipse.

The main or top pages of *EclipseWise.com* are:

> Home Page (both solar and lunar eclipses): *www.eclipsewise.com/eclipse.html*
> Solar Eclipses Page: *www.eclipsewise.com/solar/solar.html*
> Lunar Eclipses Page: *www.eclipsewise.com/lunar/lunar.html*

7 – EclipseWise.com and the 2023 Annular Eclipse

EclipseWise.com has a series of pages and resources devoted to the 2023 eclipse. The main page is located at:

> *www.eclipsewise.com/solar/SEprime/2001-2100/SE2023Oct14Aprime.html*

It provides links to detailed eclipse path maps, tables of eclipse path coordinates, Besselian elements, and more. The link to an interactive Google Map with the eclipse path plotted allows the user to zoom into an part of the path. Click on any point on the map to display the eclipse circumstances and duration of annularity at that location. This web site will continue to add features as the eclipse approaches.

The *2023 Annular Solar Eclipse Circumstances Calculator* is an interactive web page that can quickly calculate the local circumstances for the eclipse from any geographic location. The *Calculator* is located at:

> *www.EclipseWise.com/solar/SEcirc/2001-2100/SE2023Oct14Acirc.html*

An entire annular eclipse is captured in the above composite of seven separate images.
This sequence was captured during the annular solar eclipse of October 3, 2005 from Carrascosa del Campo, Spain.
© 2005 F. Espenak, www.MrEclipse.com

8 – Total and Annular Eclipses in the USA and Canada: 2001 – 2050

During the first 50 years of the 21st Century, the paths of four total and five annular solar eclipses pass through parts of the USA and Canada. The following map illustrates exactly where each of these rare celestial events will be visible from. The blue shaded paths are total eclipses, while the yellow shaded paths are annular eclipses.

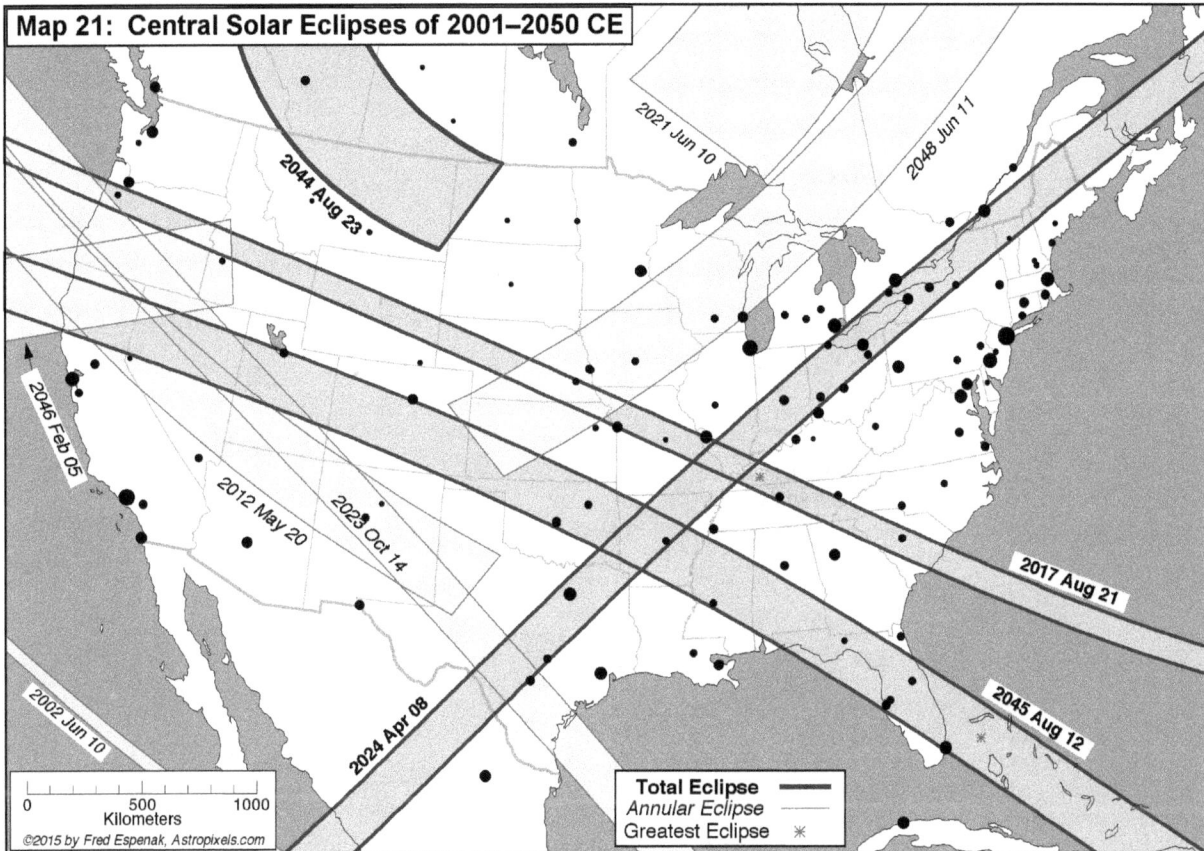

Map courtesy of *"Atlas of Central Solar Eclipses in the USA"* by Fred Espenak (2017).

The dates of these central eclipses are listed in the following table.

Total Eclipses of the Sun

1) 2017 August 17
2) 2024 April 08
3) 2044 August 23
4) 2045 August 12

Annular Eclipses of the Sun

1) 2012 May 20
2) 2021 June 10 (Canada only)
3) 2023 October 14
4) 2046 February 05
5) 2048 June 11

For more information on central eclipses in the USA, see: *www.eclipsewise.com/solar/SEcountry/SEinUSA.html*

9 – Eclipse Predictions

The algorithms and software for the eclipse predictions were developed primarily from the *Explanatory Supplement to the Astronomical Ephemeris* (Her Majesty's Nautical Almanac Office, 1974) with additional algorithms from *Elements of Solar Eclipses: 1951–2200* (Meeus, 1989). The solar and lunar ephemerides were generated from the JPL DE405. All eclipse calculations were made using a value for the Moon's radius of k=0.2722810 for the path of annularity. Center of mass coordinates for the Moon have been used without correction to the lunar limb profile. A value for ΔT of 71.3 seconds was used to convert the predictions from Terrestrial Dynamical Time to Universal Time (UT1).

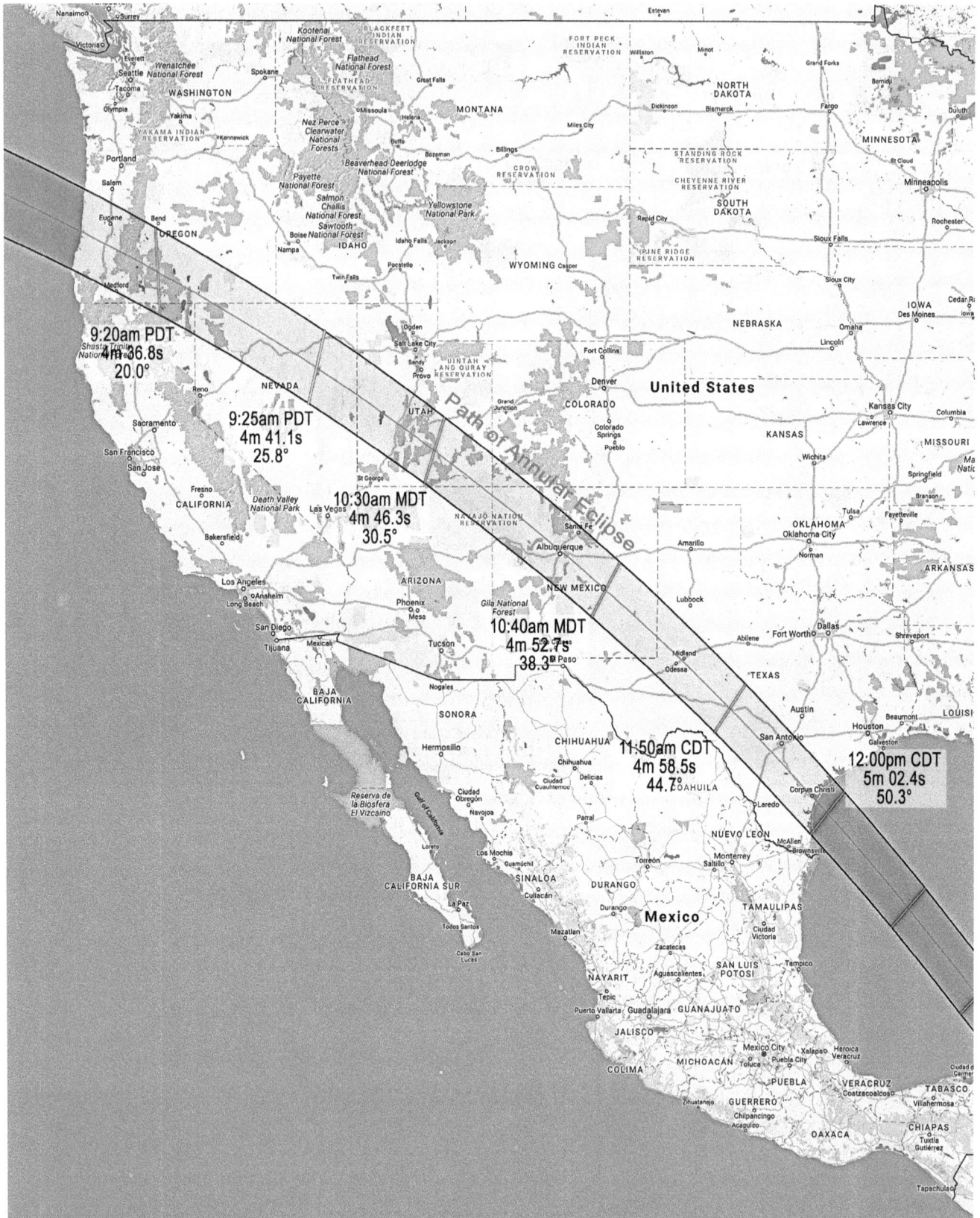

9:20am PDT
4m 36.8s
20.0°

9:25am PDT
4m 41.1s
25.8°

10:30am MDT
4m 46.3s
30.5°

10:40am MDT
4m 52.7s
38.3°

11:50am CDT
4m 58.5s
44.7°

12:00pm CDT
5m 02.4s
50.3°

Path of Annular Eclipse

Overview Map 1:
Path of 2023 Annular Solar Eclipse
Through Western USA

©2018 F. Espenak

Map data ©2018 Google, INEGI

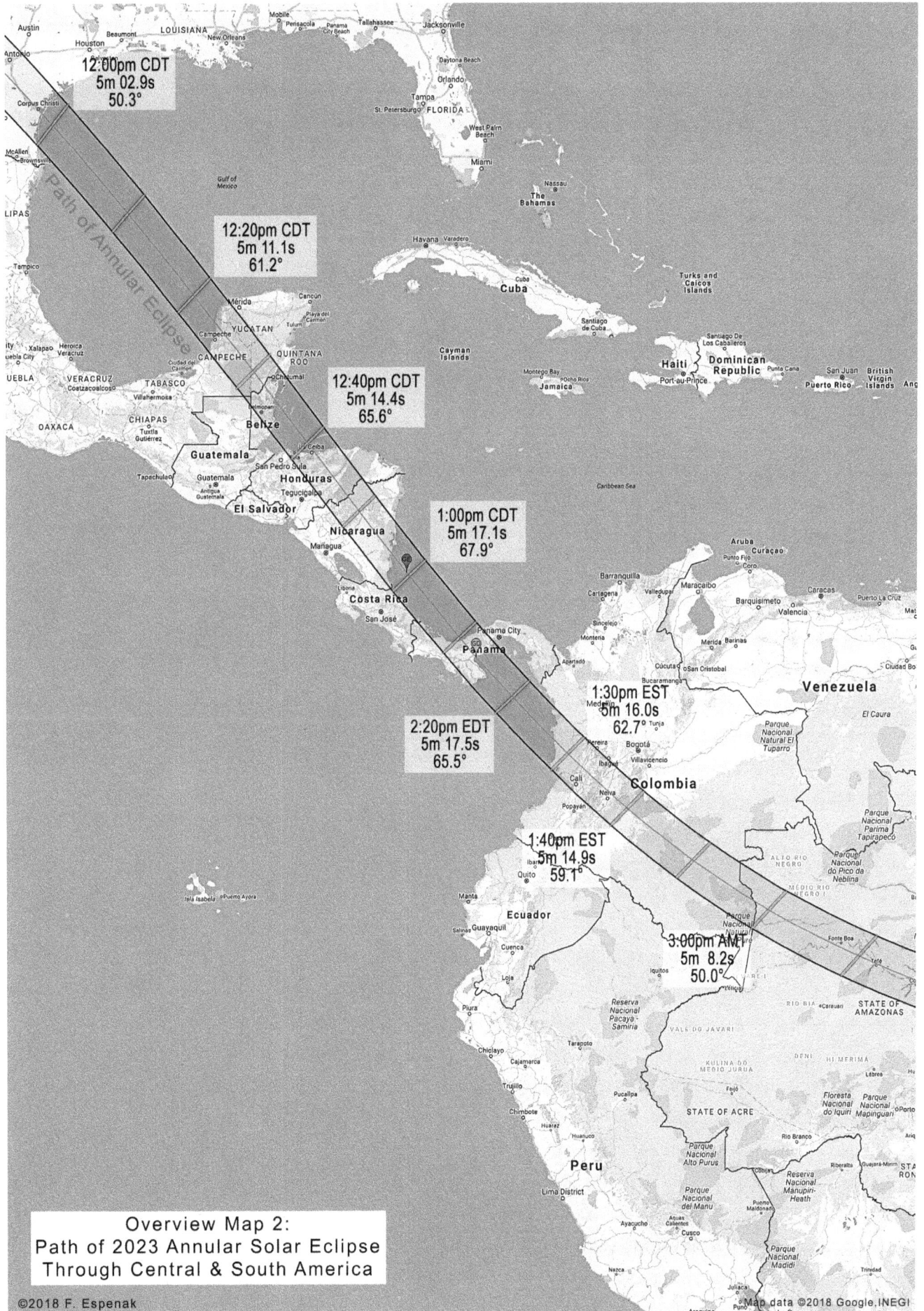

12:00pm CDT
5m 02.9s
50.3°

12:20pm CDT
5m 11.1s
61.2°

12:40pm CDT
5m 14.4s
65.6°

1:00pm CDT
5m 17.1s
67.9°

1:30pm EST
5m 16.0s
62.7°

2:20pm EDT
5m 17.5s
65.5°

1:40pm EST
5m 14.9s
59.1°

3:00pm AMT
5m 8.2s
50.0°

Overview Map 2:
Path of 2023 Annular Solar Eclipse
Through Central & South America

©2018 F. Espenak

Map data ©2018 Google, INEGI

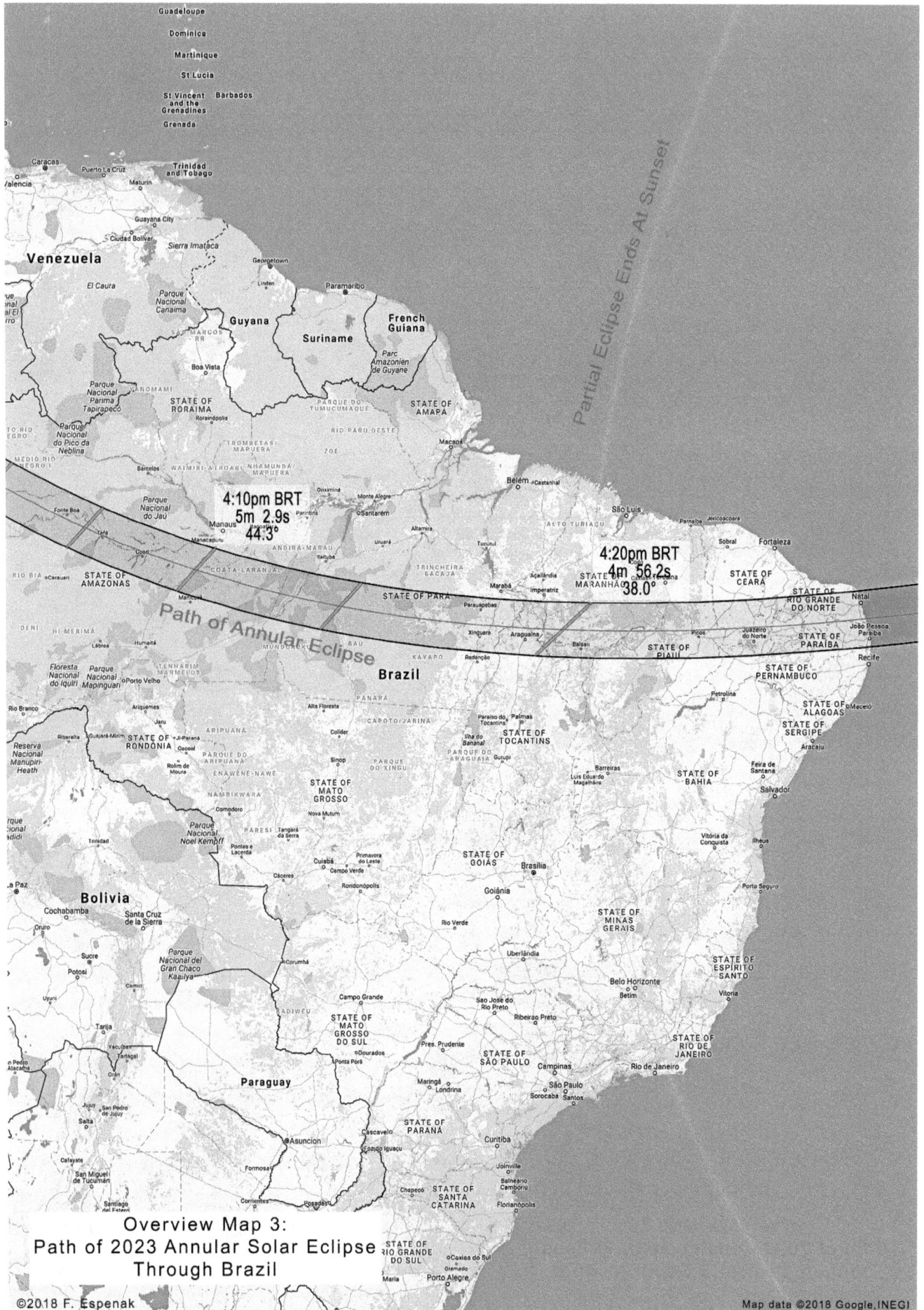

Partial Eclipse Ends At Sunset

Path of Annular Eclipse

4:10pm BRT
5m 2.9s
44.3°

4:20pm BRT
4m 56.2s
38.0°

Venezuela

Guyana

Suriname

French Guiana

Brazil

Bolivia

Paraguay

Overview Map 3:
Path of 2023 Annular Solar Eclipse
Through Brazil

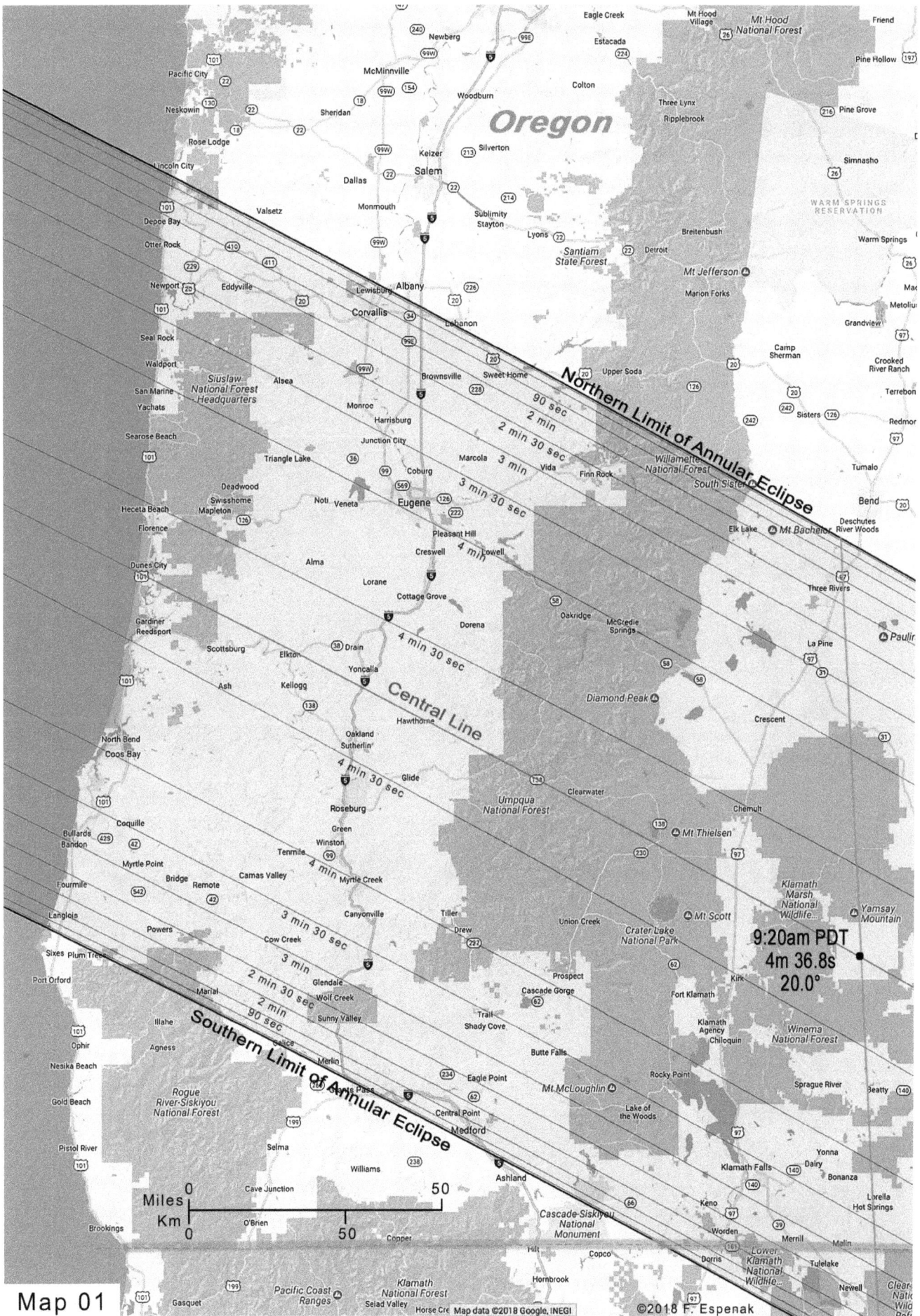

Map 01

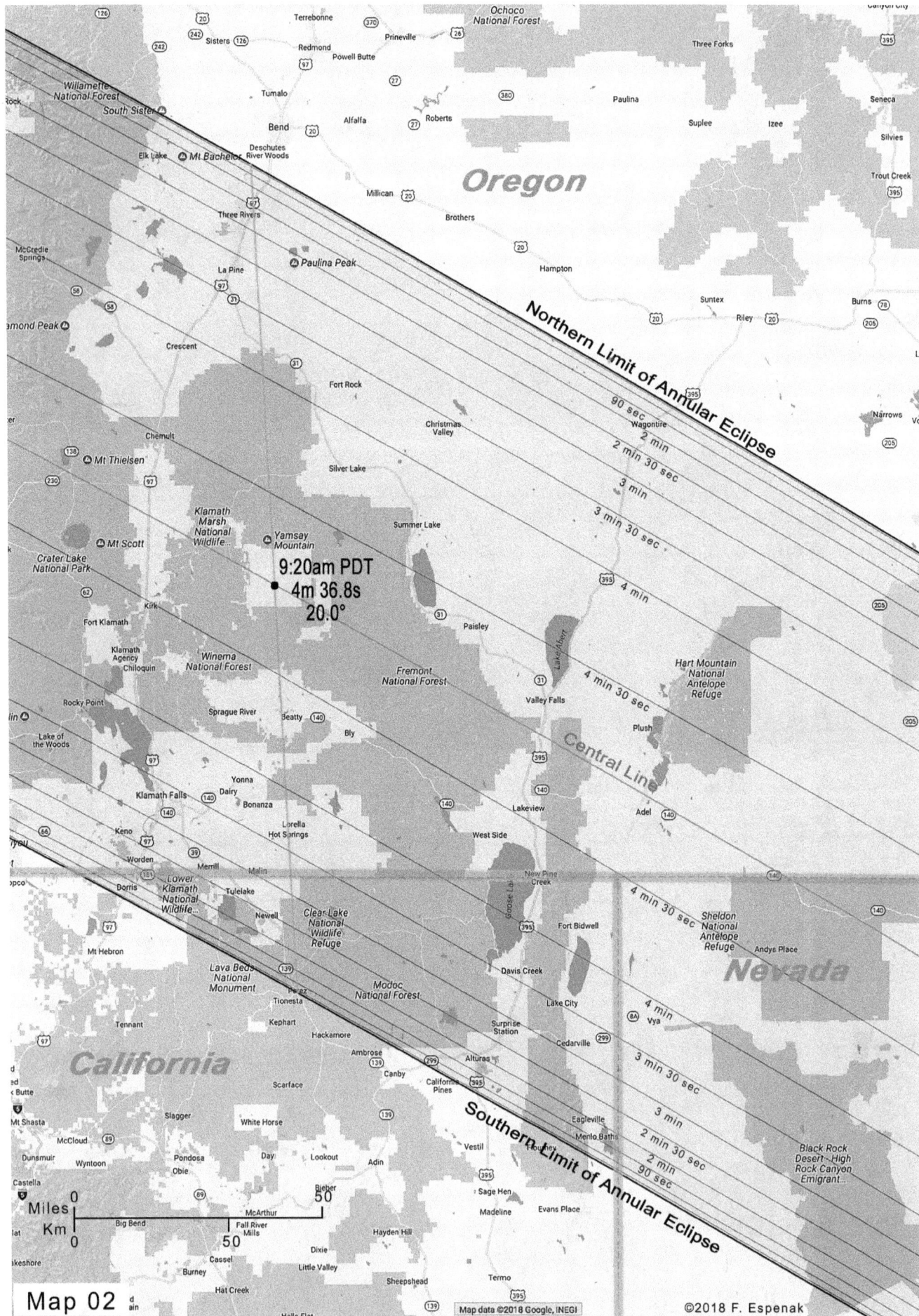

Map 02

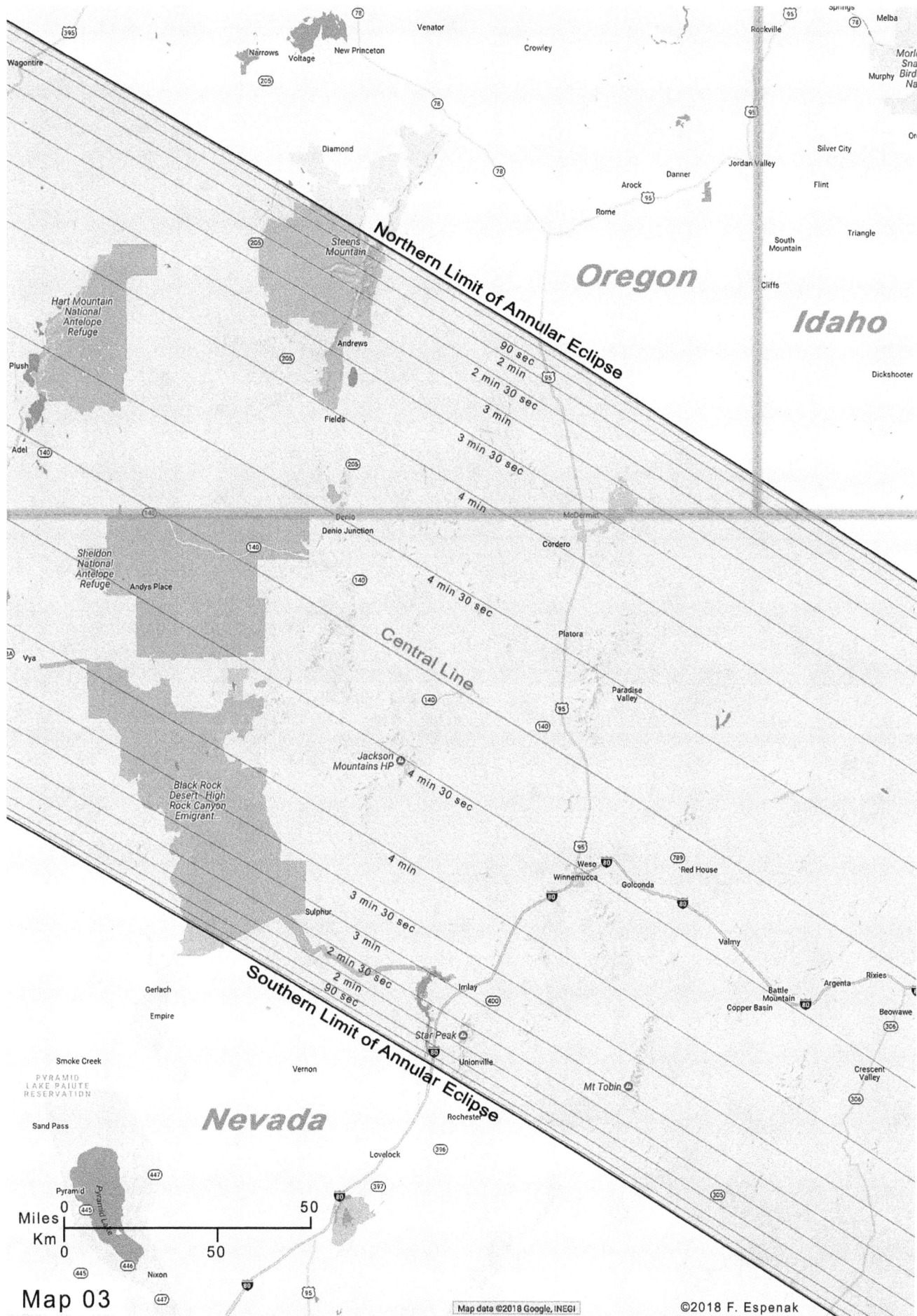

Map 03

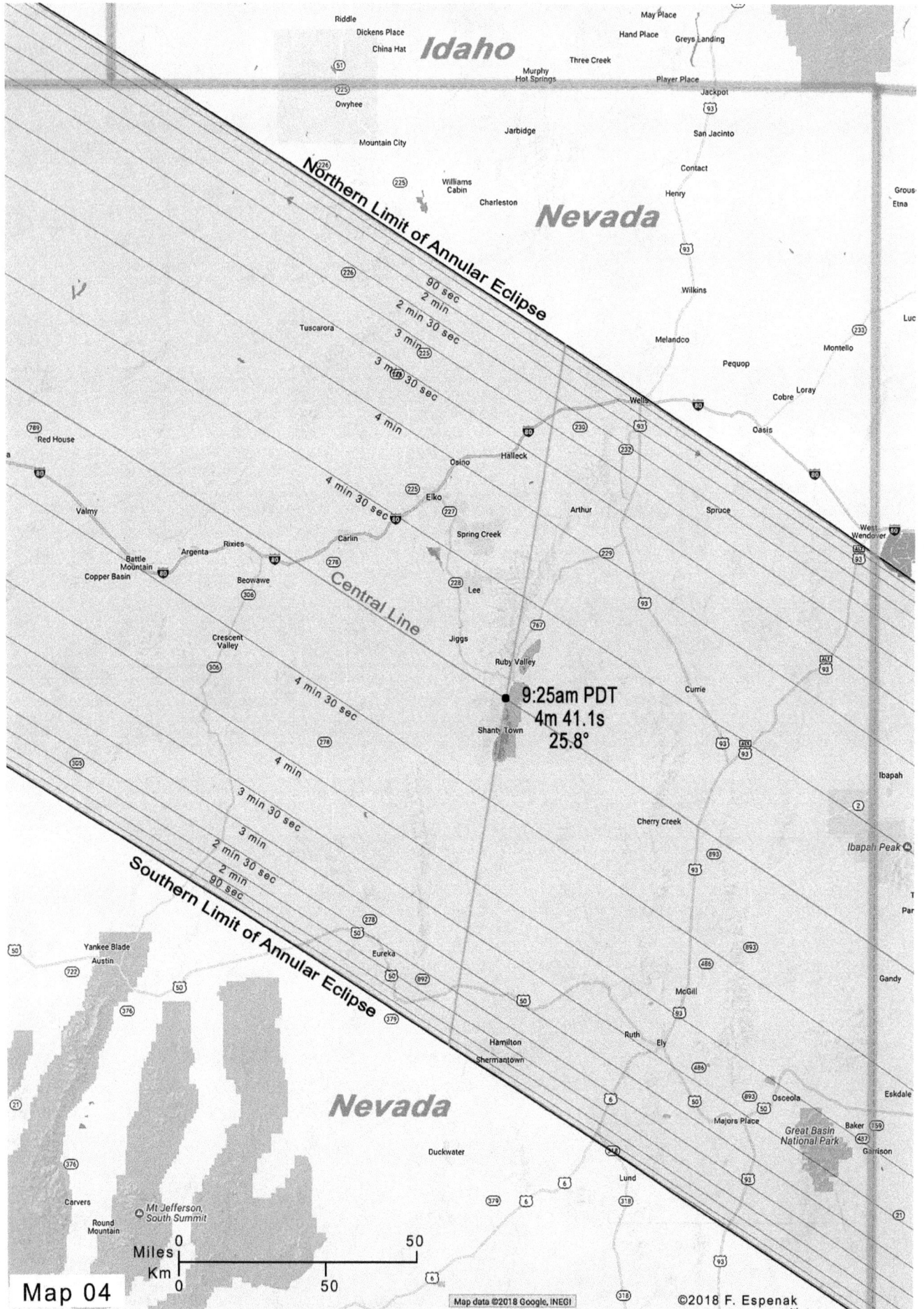

Map 04

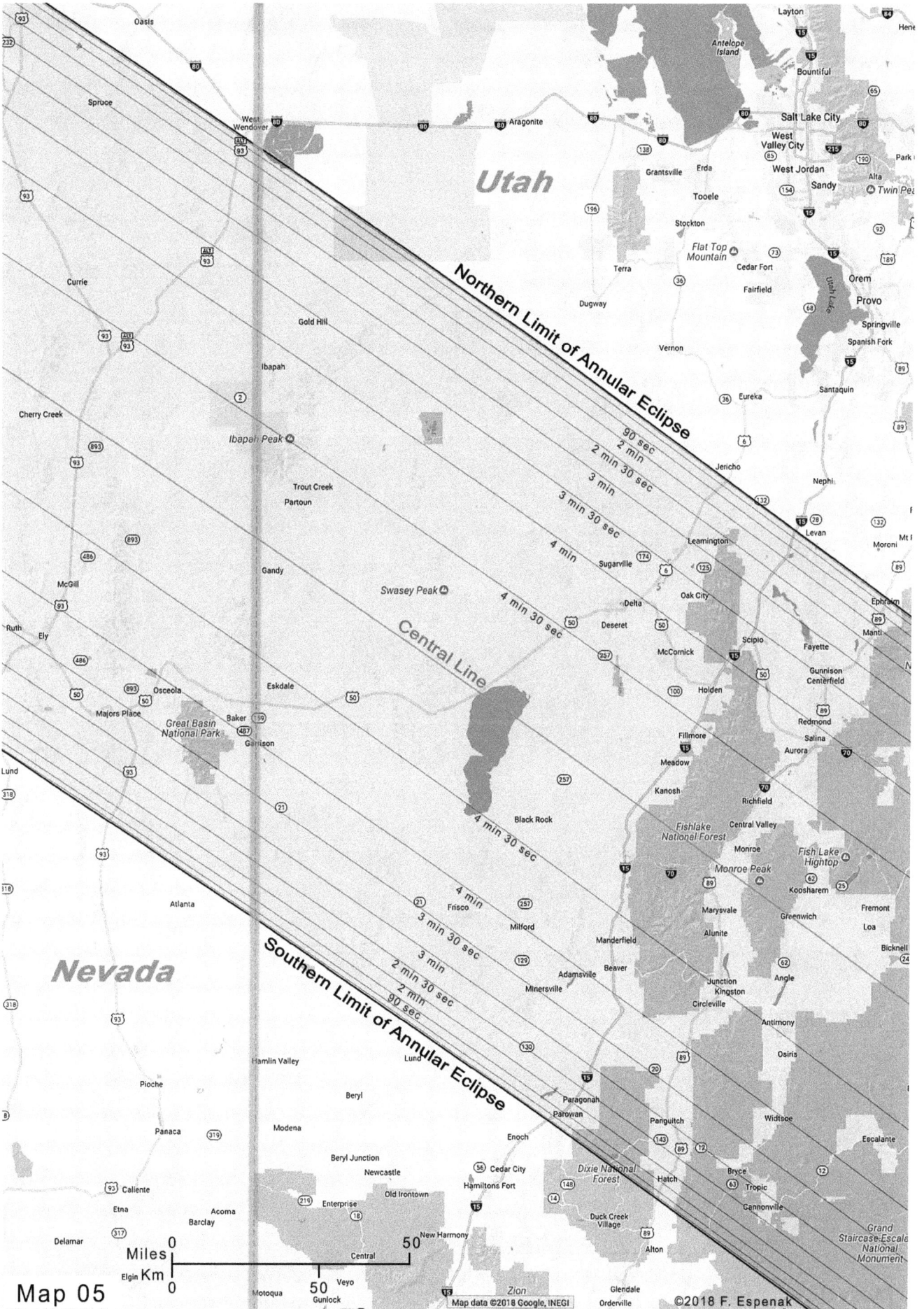

Map 05

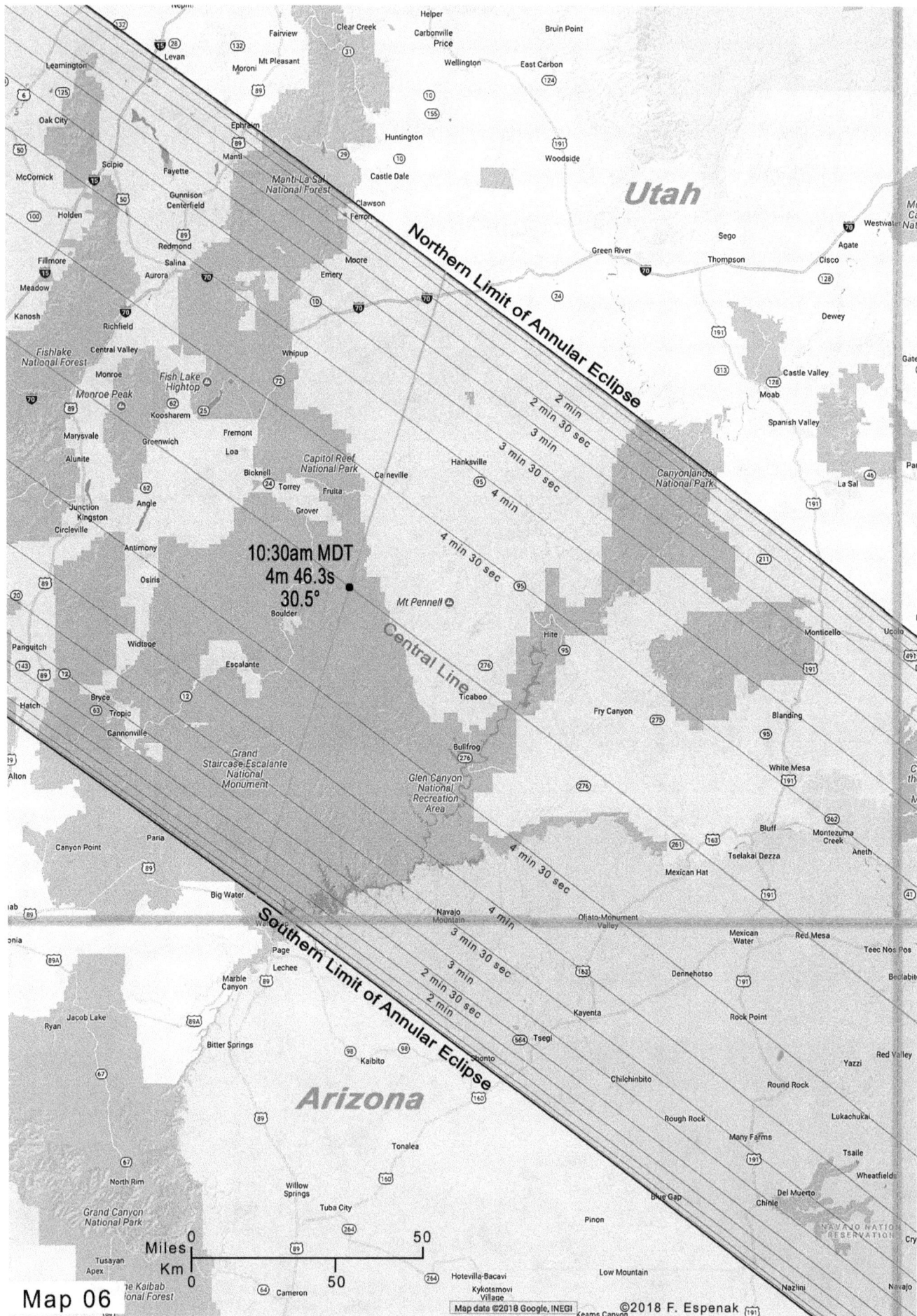

Map 06

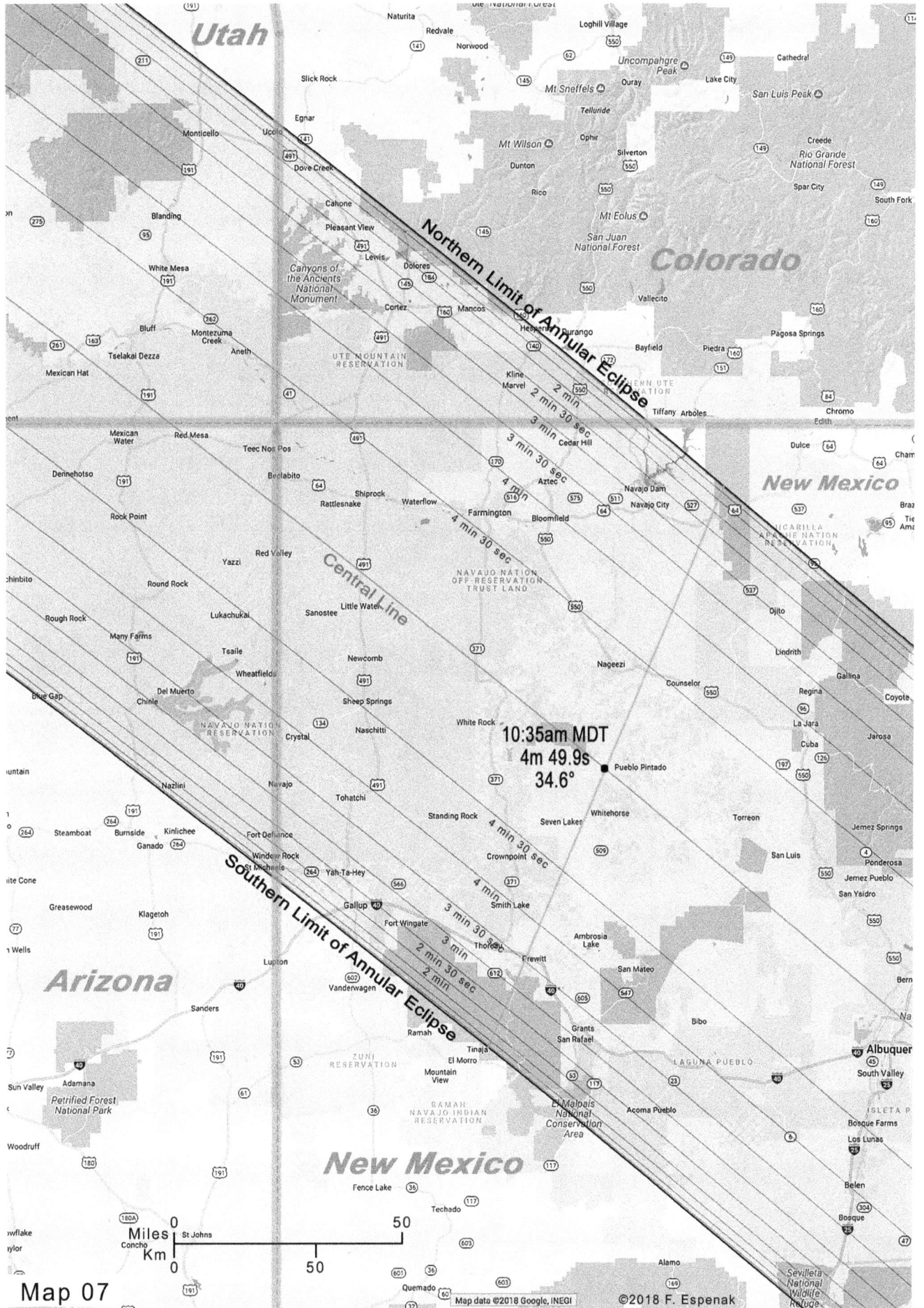

Map 07

10:35am MDT
4m 49.9s
34.6°

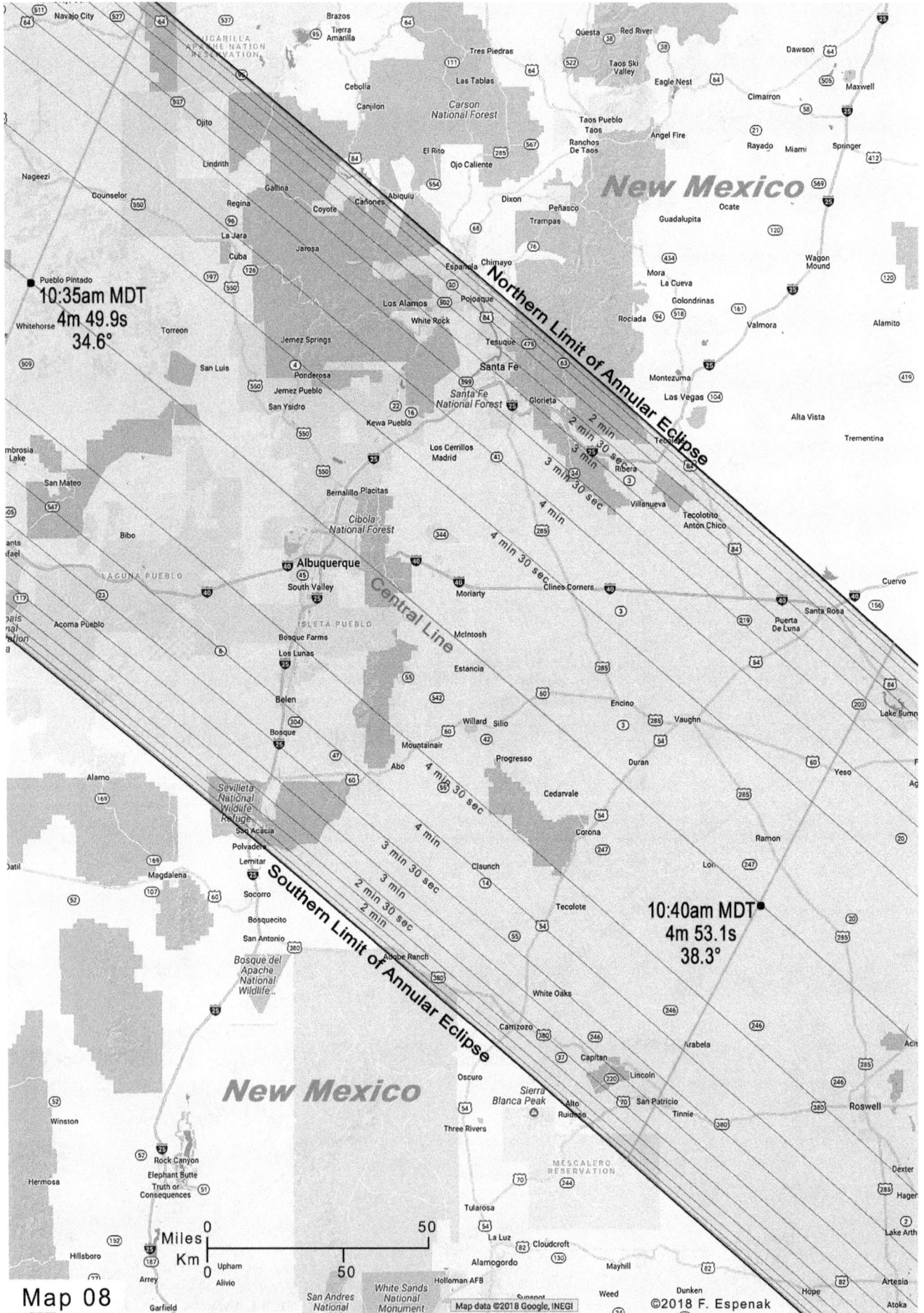

Map 08

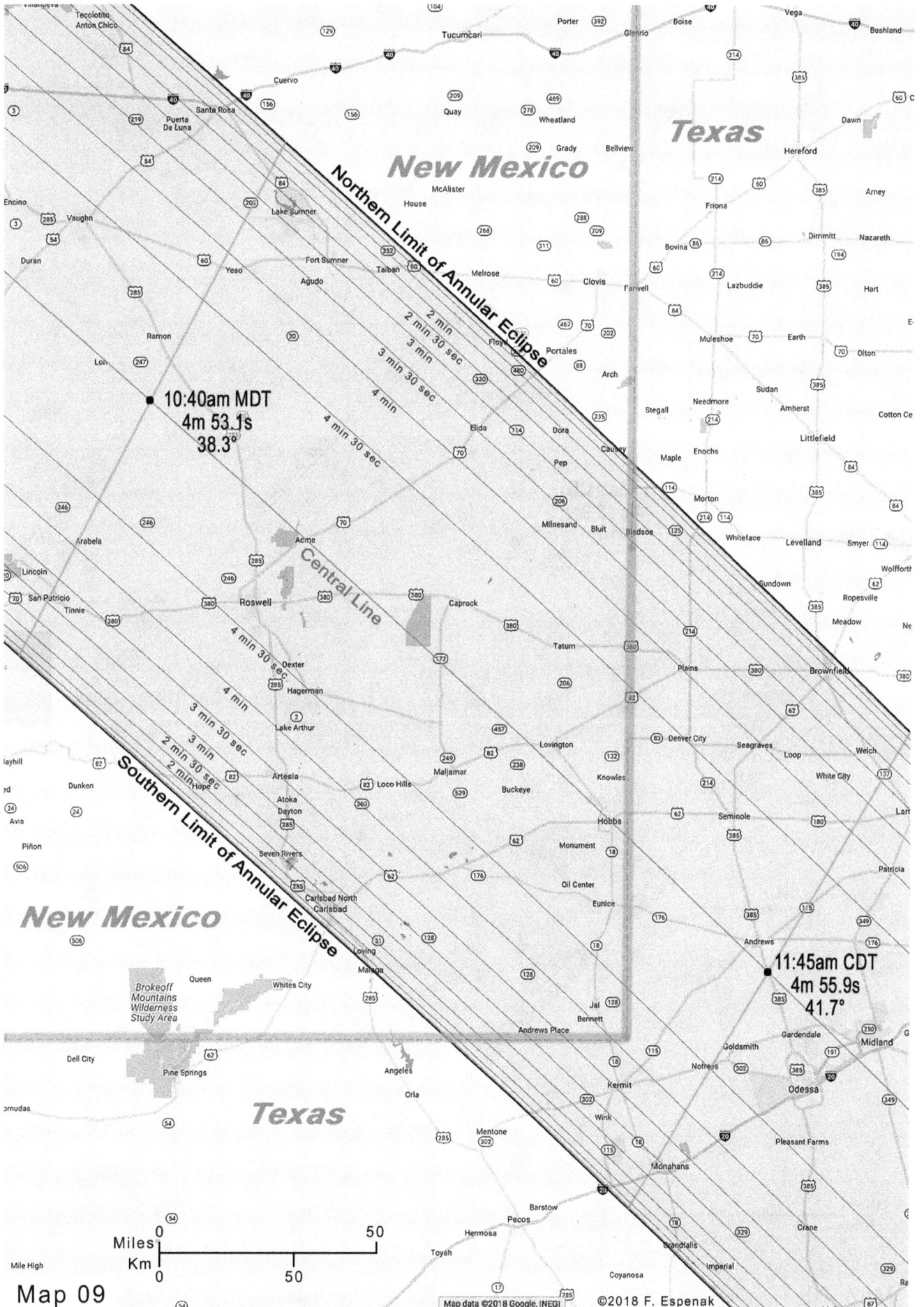

Map 09

10:40am MDT
4m 53.1s
38.3°

11:45am CDT
4m 55.9s
41.7°

Northern Limit of Annular Eclipse

Southern Limit of Annular Eclipse

Central Line

2 min
2 min 30 sec
3 min
3 min 30 sec
4 min
4 min 30 sec

4 min 30 sec
4 min
3 min 30 sec
3 min
2 min 30 sec
2 min

New Mexico

Texas

New Mexico

Texas

Miles
Km

Map data ©2018 Google, INEGI   ©2018 F. Espenak

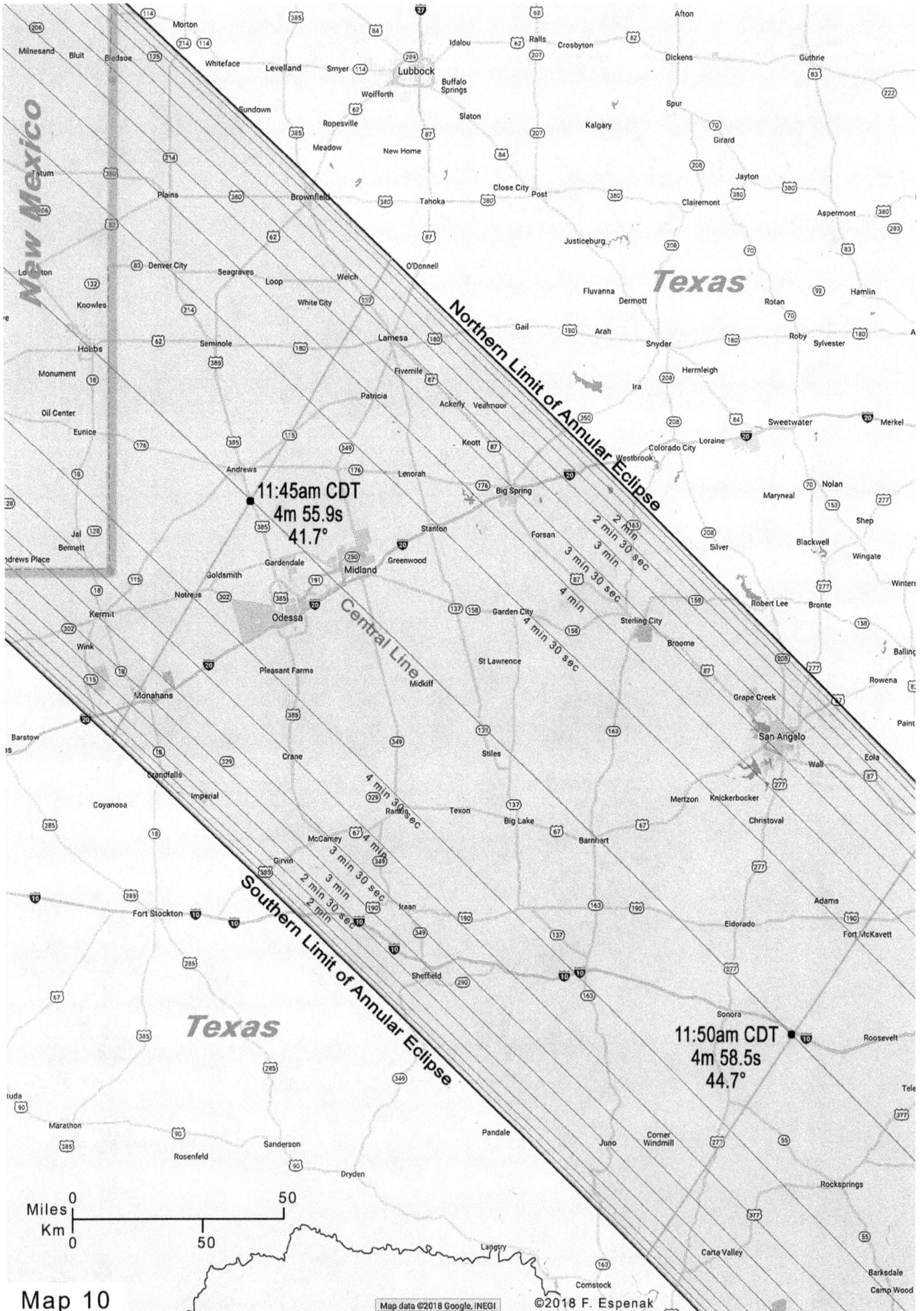

New Mexico

Texas

Northern Limit of Annular Eclipse

Central Line

Southern Limit of Annular Eclipse

Texas

11:45am CDT
4m 55.9s
41.7°

11:50am CDT
4m 58.5s
44.7°

2 min
2 min 30 sec
3 min
3 min 30 sec
4 min
4 min 30 sec

4 min 30 sec
4 min
3 min 30 sec
3 min
2 min 30 sec
2 min

Miles
0   50

Km
0   50

Map 10

Map data ©2018 Google, INEGI   ©2018 F. Espenak

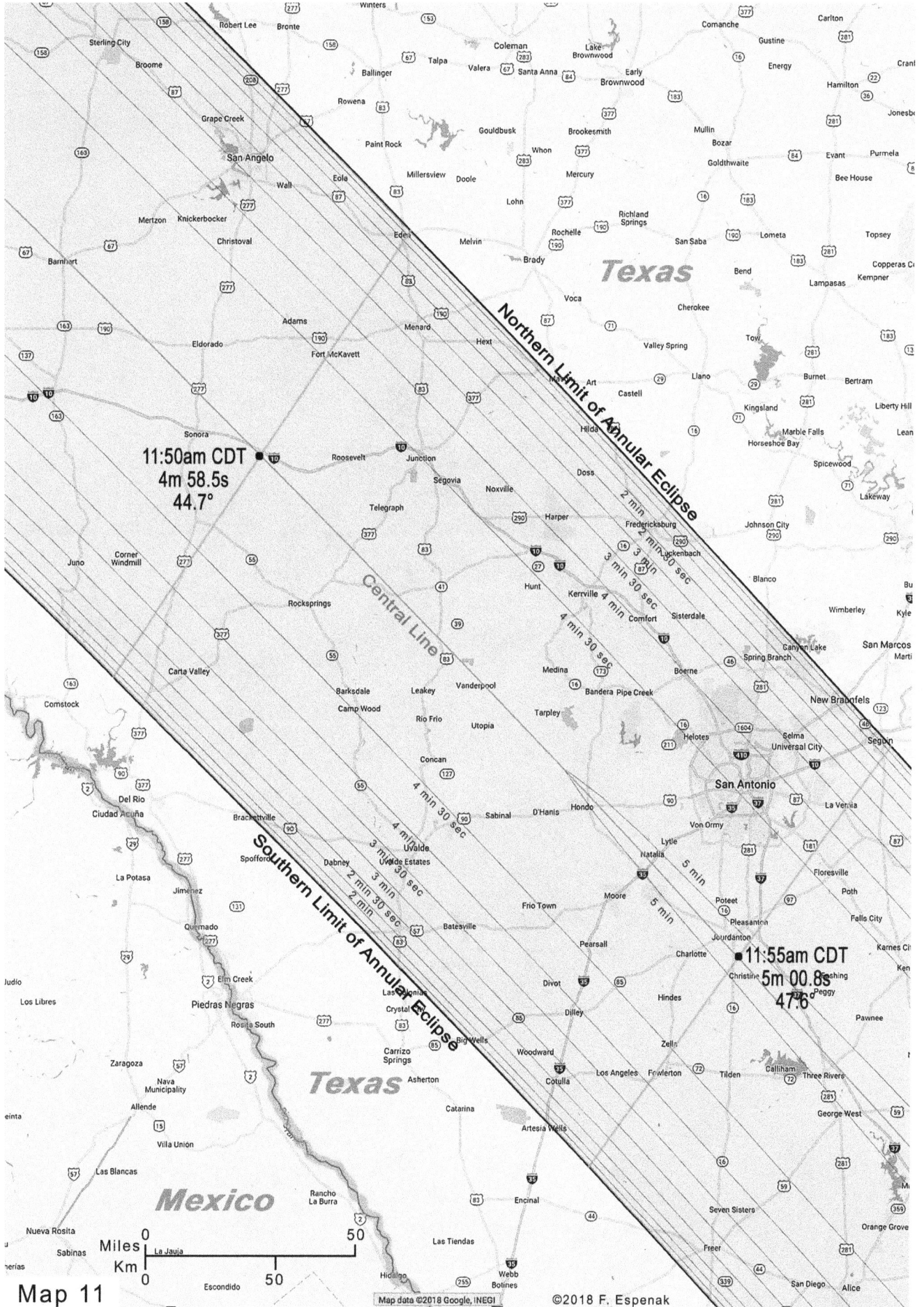

Northern Limit of Annular Eclipse

Southern Limit of Annular Eclipse

Central Line

Texas

Texas

Mexico

11:50am CDT
4m 58.5s
44.7°

11:55am CDT
5m 00.8s
47.6°

2 min
2 min 30 sec
3 min
3 min 30 sec
4 min
4 min 30 sec
5 min
5 min
4 min 30 sec
4 min
3 min 30 sec
3 min
2 min 30 sec
2 min

San Angelo
San Antonio

Miles
0   50
Km
0   50

La Jauja

Map 11

Map data ©2018 Google, INEGI   ©2018 F. Espenak

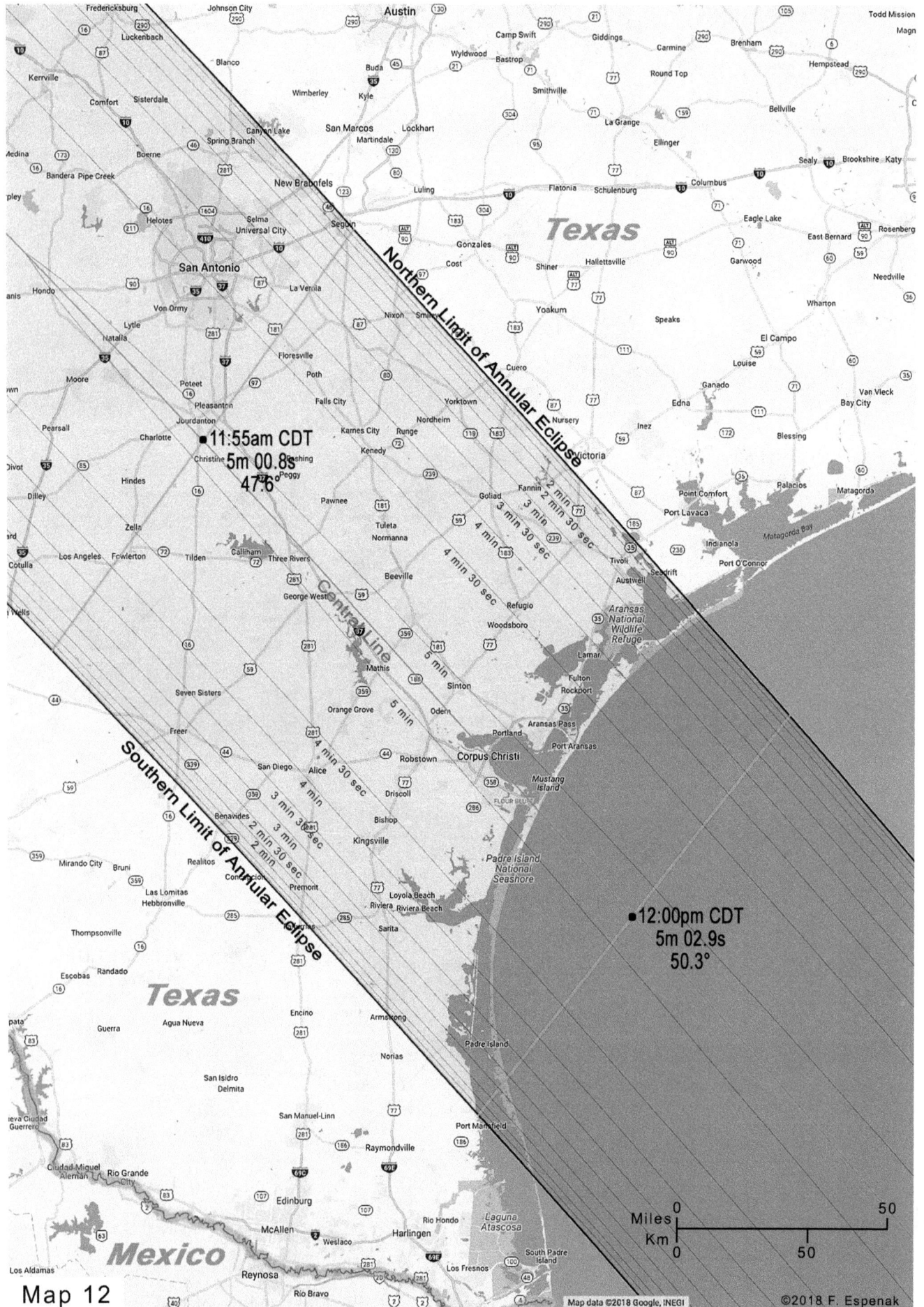

Fredericksburg Johnson City Austin Camp Swift Giddings Carmine Brenham Todd Mission Magn
Luckenbach Blanco Wyldwood Bastrop Round Top Hempstead
Kerrville Comfort Sisterdale Buda Kyle Smithville Bellville
Comfort Canyon Lake Wimberley La Grange Ellinger Sealy Brookshire Katy
Medina Boerne Spring Branch San Marcos Martindale Lockhart Schulenburg Columbus Rosenberg
Bandera Pipe Creek New Braunfels Luling Flatonia Eagle Lake East Bernard
Helotes Selma Seguin **Texas** Gonzales Garwood Needville
San Antonio Universal City La Vernia Cost Shiner Hallettsville Wharton
Hondo Nixon Smiley Yoakum Speaks El Campo Van Vleck
Lytle Von Ormy Floresville Poth Falls City Yorktown Nordheim Runge Louise Bay City
Natalia Moore Poteet Karnes City Kenedy Nursery Victoria Edna Ganado
Pearsall Charlotte Jourdanton Pawnee Cuero Inez Blessing
Christine Pleasanton Flushing Peggy Tuleta Goliad Fannin Point Comfort Palacios Matagorda
Hindes Normanna Port Lavaca Indianola
Los Angeles Fowlerton Tilden Calliham Three Rivers Beeville Tivoli Seadrift Port O'Connor
Dilley Zella George West Refugio Austwell
Cotulla Seven Sisters Mathis Woodsboro Aransas National Wildlife Refuge
Freer Orange Grove Sinton Odem Lamar Fulton Rockport
San Diego Alice Robstown Portland Aransas Pass Port Aransas
Driscoll Corpus Christi Mustang Island
Benavides Bishop Flour Bluff
Mirando City Bruni Realitos Concepcion Premont Kingsville
Las Lomitas Hebbronville Loyola Beach Riviera Riviera Beach Padre Island National Seashore
Thompsonville Sarita
Escobas Randado **Texas** Encino Armstrong
Guerra Agua Nueva Norias Padre Island
San Isidro Delmita
San Manuel-Linn Port Mansfield
Nueva Ciudad Guerrero Raymondville
Ciudad Miguel Aleman Rio Grande City Edinburg Rio Hondo Laguna Atascosa
McAllen Weslaco Harlingen South Padre Island
Mexico Reynosa Rio Bravo Los Fresnos
Los Aldamas

•11:55am CDT
5m 00.8s
47.6°

•12:00pm CDT
5m 02.9s
50.3°

Northern Limit of Annular Eclipse
Southern Limit of Annular Eclipse
Central Line

2 min 2 min 30 sec 3 min 3 min 30 sec 4 min 4 min 30 sec 5 min

Miles 0 50
Km 0 50

Map 12

Map data ©2018 Google, INEGI
©2018 F. Espenak

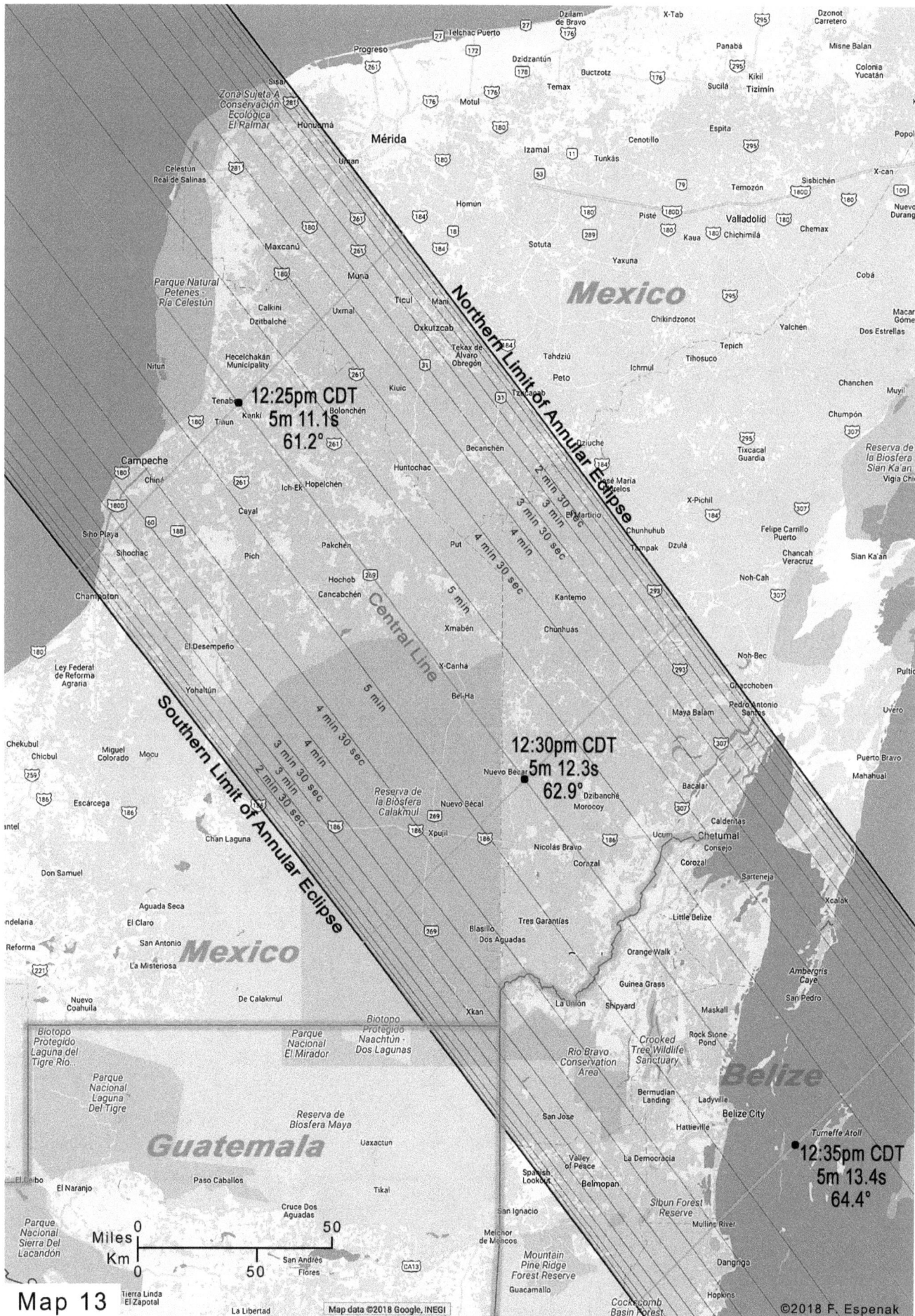

Map 13

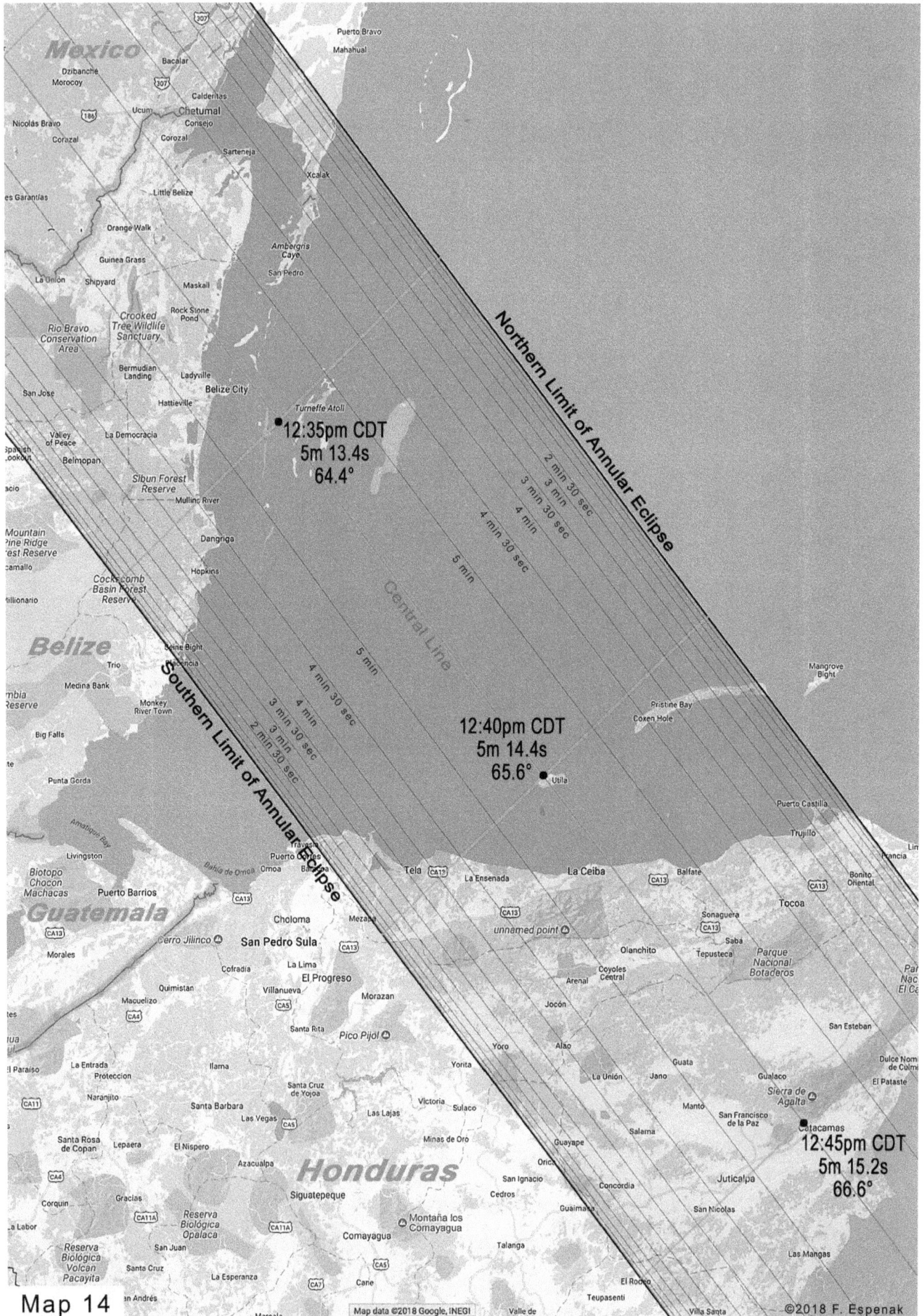

Map 14

12:35pm CDT
5m 13.4s
64.4°

12:40pm CDT
5m 14.4s
65.6°

12:45pm CDT
5m 15.2s
66.6°

Northern Limit of Annular Eclipse

Southern Limit of Annular Eclipse

Central Line

©2018 F. Espenak

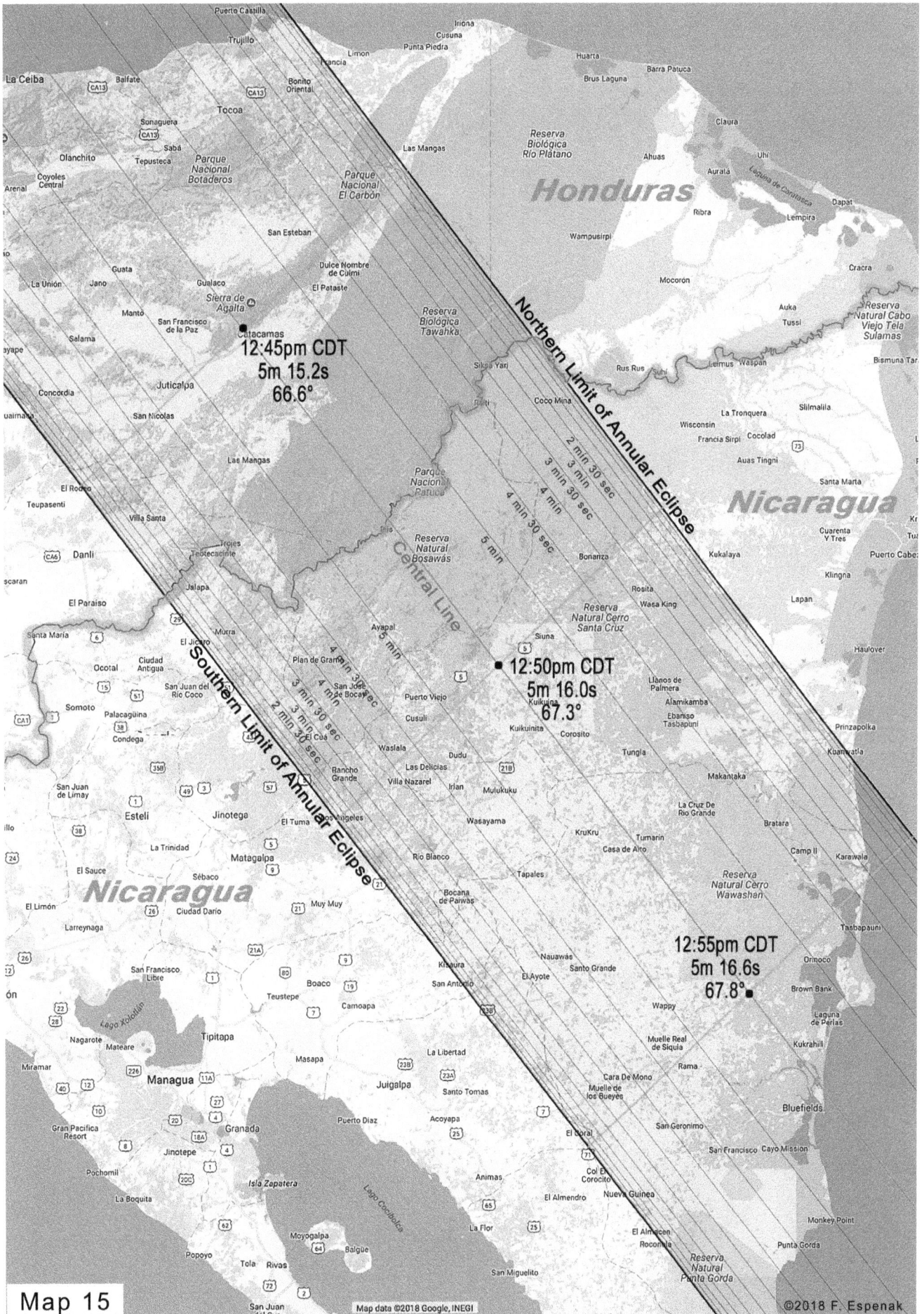

Honduras

Nicaragua

Nicaragua

Northern Limit of Annular Eclipse

Southern Limit of Annular Eclipse

Central Line

2 min 30 sec
3 min
3 min 30 sec
4 min
4 min 30 sec
5 min

12:45pm CDT
5m 15.2s
66.6°

12:50pm CDT
5m 16.0s
67.3°

12:55pm CDT
5m 16.6s
67.8°

Puerto Castilla
Trujillo
Iriona
Cusuna
Punta Piedra
Limon
Francia
Bonito Oriental
Huarta
Brus Laguna
Barra Patuca
La Ceiba
Balfate
CA13
Tocoa
CA13
Las Mangas
Reserva Biológica Río Plátano
Claura
Ahuas
Auratá
Uhi
Sonaguera
CA13
Sabá
Tepusteca
Olanchito
Parque Nacional Botaderos
Parque Nacional El Carbón
Dapat
Mocorón
Ribra
Lempira
Coyoles Central
Arenal
San Esteban
Wampusirpi
Cracra
Auka
Tussi
Reserva Natural Cabo Viejo Tela Sulamas
ayape
Guata
Jano
Gualaco
Dulce Nombre de Culmi
El Patate
La Unión
Mantó
Sierra de Agalta
San Francisco de la Paz
Catacamas
Reserva Biológica Tawahka
Rus Rus
Uhi
Leimus
Waspan
Bismuna Tar.
Salama
Siksa Yari
Coco Mina
La Tronquera
Slilmalila
Concordia
Juticalpa
Riati
Wisconsin
Francia Sirpi
Cocolad
73
uaimara
San Nicolas
Auas Tingni
Santa Marta
Las Mangas
Parque Nacional Patuca
Bonanza
Cuarenta Y Tres
Tu
El Rodeo
Teupasenti
Reserva Natural Bosawás
Rosita
Wasa King
Kukalaya
Puerto Cabe
Villa Santa
Rli
Reserva Natural Cerro Santa Cruz
Klingna
CA6
Danli
Troles
Teotecacinte
Siuna
Lapan
Haulover
El Paraiso
Jalapa
Ayapal
5 min
El Jicaro
Murra
Plan de Grama
San José de Bocay
Siuna
S
Kuikuina
Llanos de Palmera
Alamikamba
29
Santa María
6
Ocotal
Ciudad Antigua
San Juan del Río Coco
Puerto Viejo
Cusuli
Kuikuinita
Corosito
Ebaniso Tasbapuni
Prinzapolka
15
51
Somoto
Palacagüina
Condega
38
Waslala
Dudu
21B
Tungla
Makantaka
Kuamwatla
CA1
3
35B
Las Delicias
Villa Nazarel
Irlan
Mulukuku
La Cruz De Río Grande
Bratara
San Juan de Limay
49
3
57
Rancho Grande
Esteli
Jinotega
El Tuma
Los Angeles
Wasayama
KruKru
Tumarin
Casa de Alto
Camp II
Karawala
38
La Trinidad
Río Blanco
Reserva Natural Cerro Wawashan
24
El Sauce
Sébaco
Matagalpa
9
Tapales
Tasbapauni
26
Ciudad Darío
Muy Muy
Bocana de Paiwas
El Limón
Larreynaga
21A
Kisaura
Nauawás
Santo Grande
Orinoco
Brown Bank
12
26
San Francisco Libre
1
80
Boaco
19
San Antonio
El Ayote
Wappy
Laguna de Perlas
22
28
Teustepe
7
Camoapa
Muelle Real de Siquia
Kukrahill
Nagarote
Mateare
Tipitapa
La Libertad
226
Masapa
23B
23A
Cara De Mono
Rama
40
12
Managua
11A
Juigalpa
Santo Tomas
Muelle de los Bueyes
San Geronimo
Bluefields
10
27
Puerto Diaz
Acoyapa
25
7
San Francisco
Cayo Mission
Gran Pacifica Resort
2D
Granada
El Coral
71
8
Jinotepe
1
Col El Corocito
Pochomil
20C
Animas
Nueva Guinea
El Almendro
Monkey Point
62
Isla Zapatera
65
La Boquita
Moyogalpa
La Flor
25
El Almacen
Roconila
Punta Gorda
Popoyo
64
Balgüe
San Miguelito
Reserva Natural Punta Gorda
Tola
Rivas
72
2
San Juan del Sur

Map 15

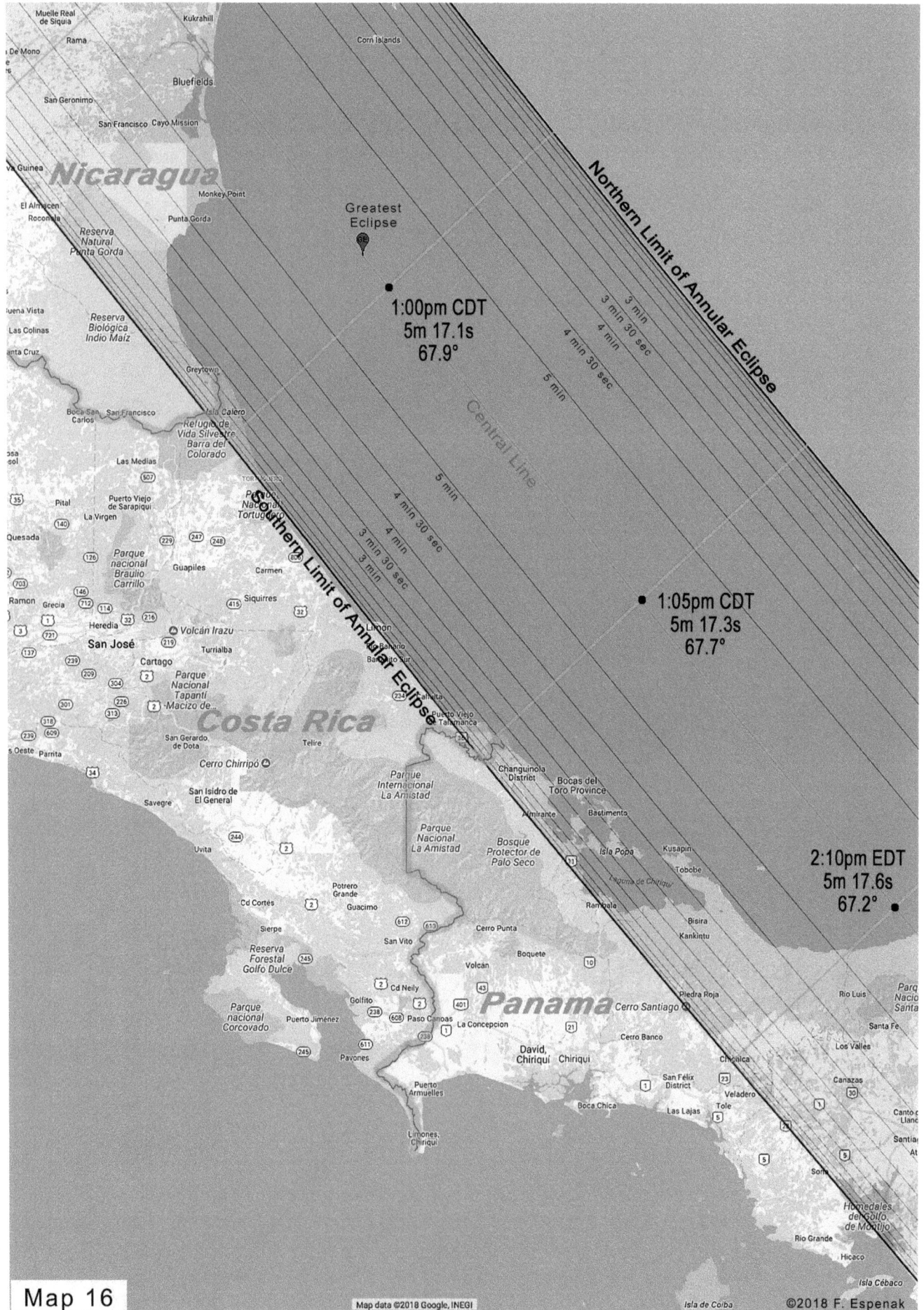

Map 16

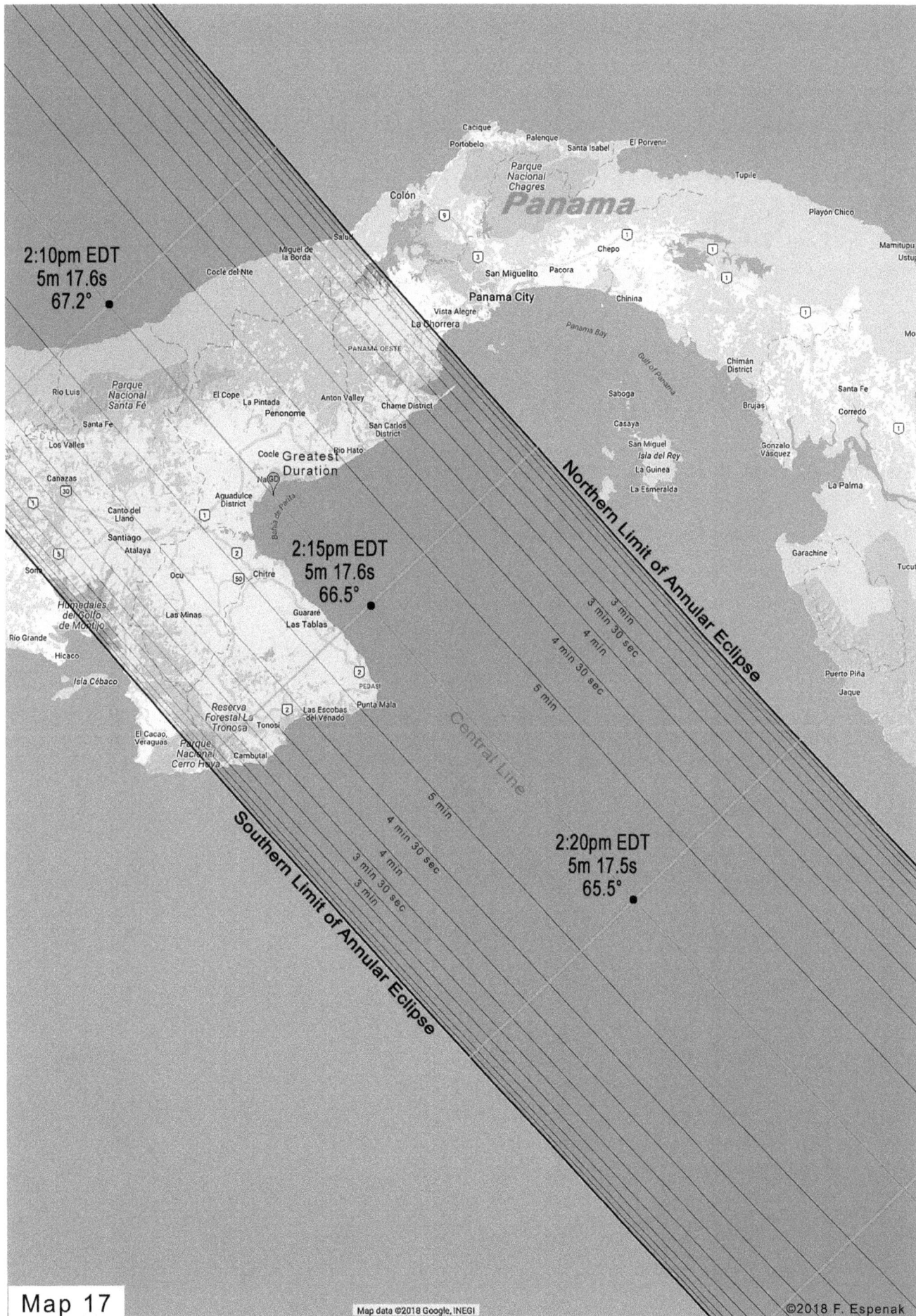

Map 17

2:10pm EDT
5m 17.6s
67.2° •

2:15pm EDT
5m 17.6s
66.5° •

2:20pm EDT
5m 17.5s
65.5° •

Cacique
Portobelo
Palenque
Santa Isabel
El Porvenir
Tupile
Parque
Nacional
Chagres
Colón
Panama
Playón Chico
Salud
Mamitupu
Ustupo
Miguel de
la Borda
Chepo
Cocle del Nte
San Miguelito
Pacora
Mort
Vista Alegre
Panama City
Chinina
La Chorrera
PANAMA OESTE
Panama Bay
Gulf of Panama
Parque
Nacional
Santa Fé
El Cope
Anton Valley
Chame District
Chimán
District
Río Luis
La Pintada
Penonome
Saboga
Santa Fe
Santa Fe
San Carlos
District
Casaya
Corredó
Los Valles
Cocle
Greatest
Duration
Río Hato
San Miguel
Isla del Rey
Brujas
Cañazas
Na(30)
Aquadulce
District
La Guinea
Gonzalo
Vásquez
La Palma
Canto del
Llano
Bahía de Parita
La Esmeralda
Santiago
Atalaya
Chitré
Guararé
Las Tablas
Garachine
Ocu
Tucuti
Las Minas
Humedales
del Golfo
de Montijo
PEDASI
Río Grande
Hicaco
Puerto Piña
Isla Cébaco
Reserva
Forestal La
Tronosa
Las Escobas
del Venado
Punta Mala
Jaque
El Cacao
Veraguas
Tonosi
Parque
Nacional
Cerro Hoya
Cambutal

Northern Limit of Annular Eclipse

3 min
3 min 30 sec
4 min
4 min 30 sec
5 min

Central Line

5 min
4 min 30 sec
4 min
3 min 30 sec
3 min

Southern Limit of Annular Eclipse

©2018 F. Espenak

Colombia

Bogotá
CHAPINERO

Medellín

Northern Limit of Annular Eclipse

Southern Limit of Annular Eclipse

Central Line

1:30pm EST
5m 16.6s
62.7°

1:35pm EST
5m 15.8s
61.0°

3 min
3 min 30 sec
4 min
4 min 30 sec
5 min
5 min
4 min 30 sec
4 min
3 min 30 sec
3 min

Map data ©2018 Google, INEGI

Map 18

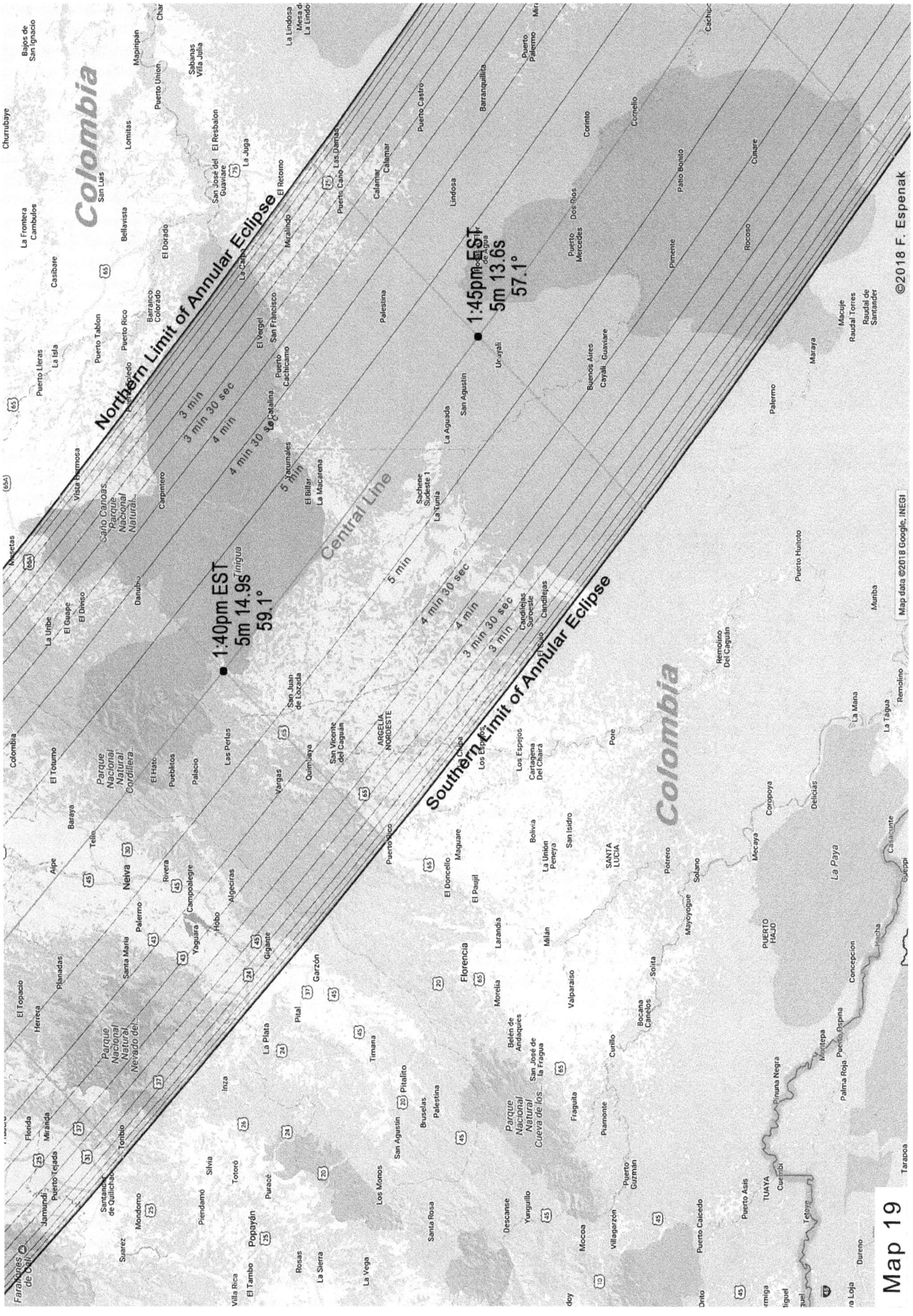

©2018 F. Espenak

Colombia

Northern Limit of Annular Eclipse

Central Line

Southern Limit of Annular Eclipse

Colombia

1:40pm EST
5m 14.9s
59.1°

1:45pm EST
5m 13.6s
57.1°

3 min
3 min 30 sec
4 min
4 min 30 sec
5 min

5 min

4 min 30 sec
4 min
3 min 30 sec
3 min

Map data ©2018 Google, INEGI

Map 19

35

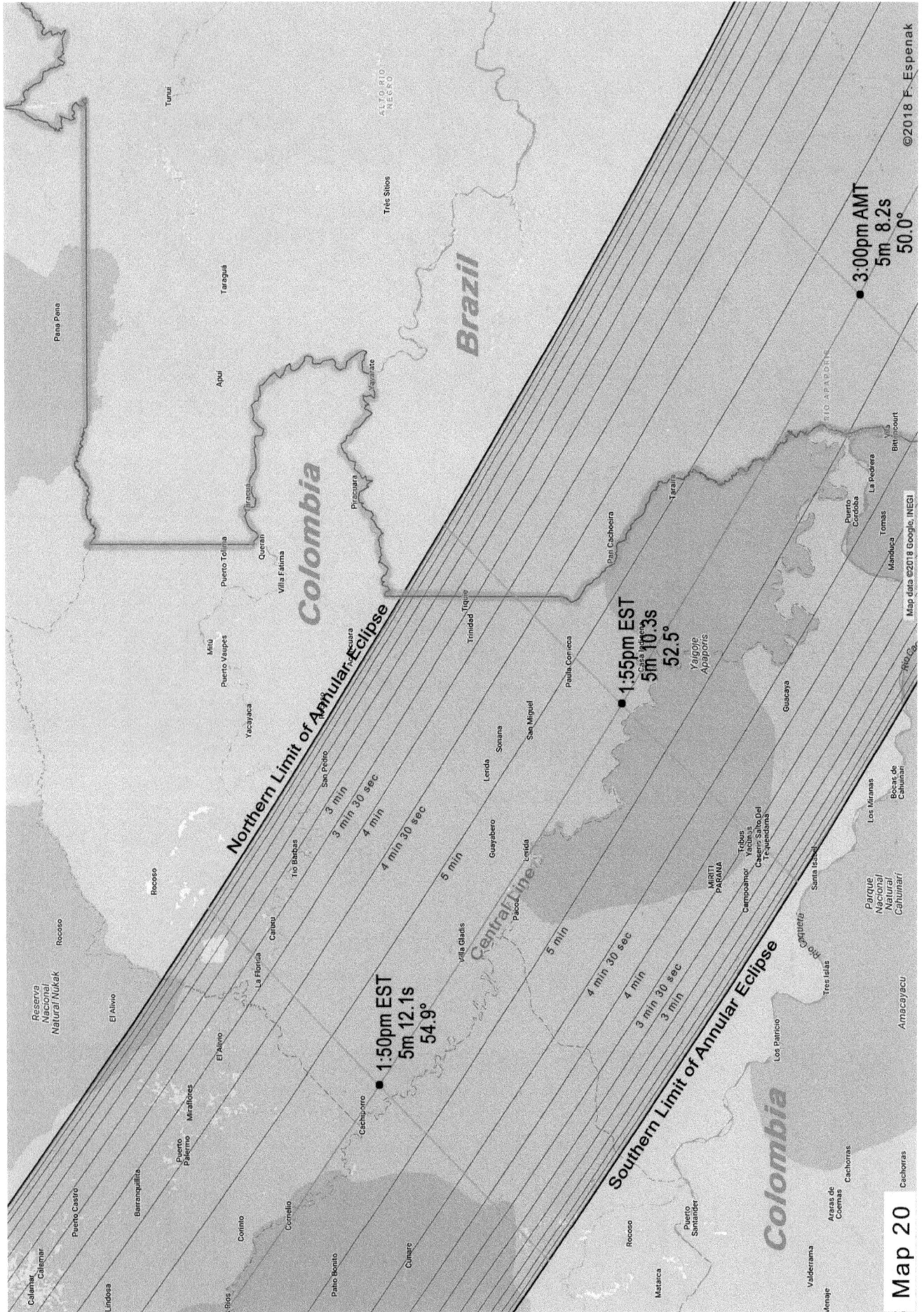

Map 20

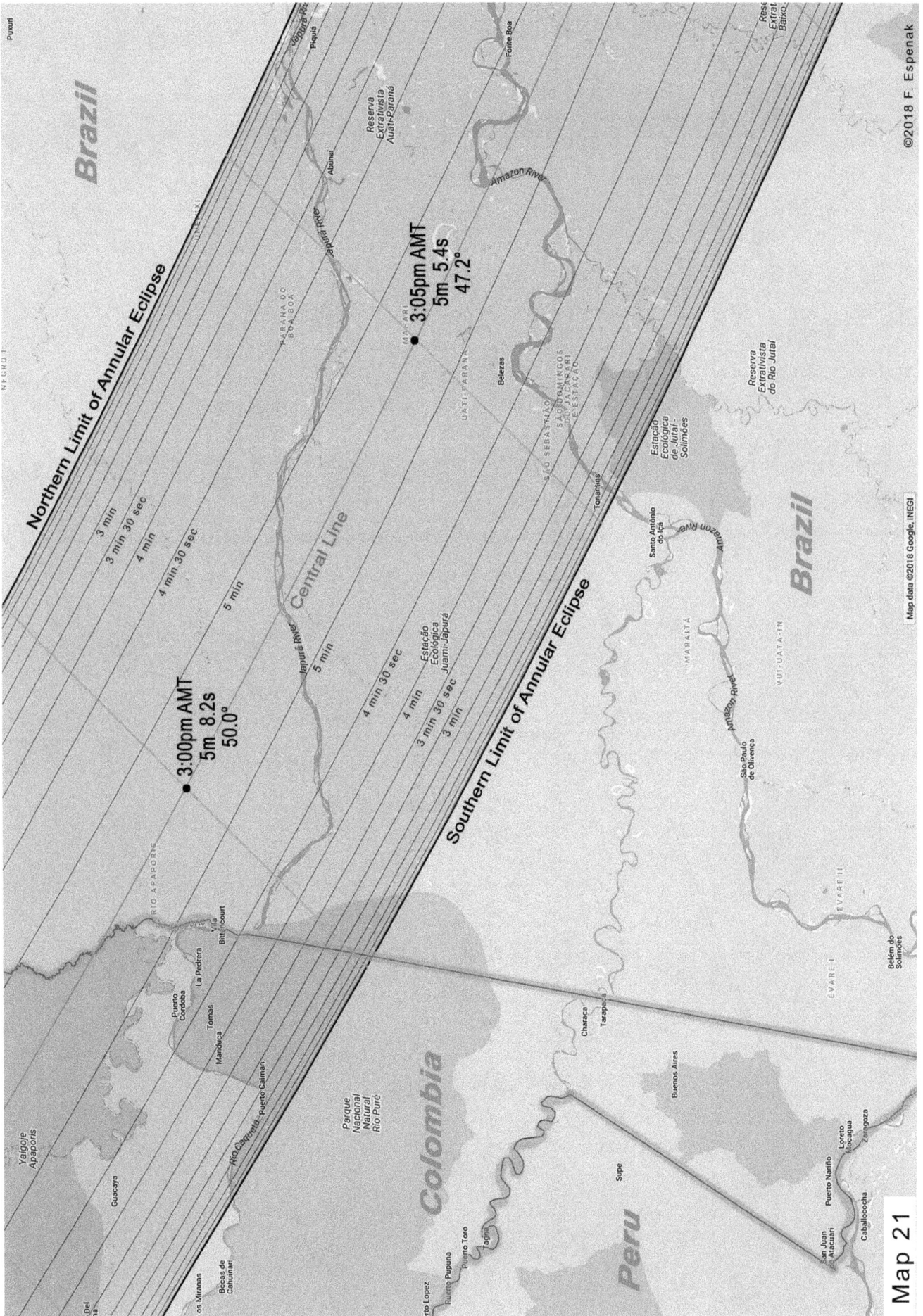

©2018 F. Espenak

Brazil

Brazil

Colombia

Peru

Puxuri

Piquiá

Fonte Boa

Reserva Extrativista Baixo...

Reserva Extrativista Auati-Paraná

Abunaí

Amazon River

Northern Limit of Annular Eclipse

3:05pm AMT
5m 5.4s
47.2°

Belezas

Reserva Extrativista do Rio Jutaí

Tonantins

Estação Ecológica de Jutaí-Solimões

3 min

3 min 30 sec

4 min

4 min 30 sec

Central Line

5 min

Santo Antônio do Içá

Amazon River

MARAITÁ

5 min

Estação Ecológica Juami-Japurá

4 min 30 sec

4 min

3 min 30 sec

3 min

Southern Limit of Annular Eclipse

Bittencourt

La Pedrera

Puerto Córdoba

Tomás

Mariduca

Rio Caquetá-Puerto Calunán

Characa

Tarapacá

São Paulo de Olivença

EVARE II

Belém do Solimões

Buenos Aires

3:00pm AMT
5m 8.2s
50.0°

Yaigojé Apaporis

Guacayá

Parque Nacional Natural Río Puré

Supe

Puerto Nariño

Loreto
Yacucaya

Zaragoza

Los Mirones

Bocas de Cahuinarí

Puerto Toro

Pupuna

San Juan de Atacuari

Caballococha

Map data ©2018 Google, INEGI

Map 21

37

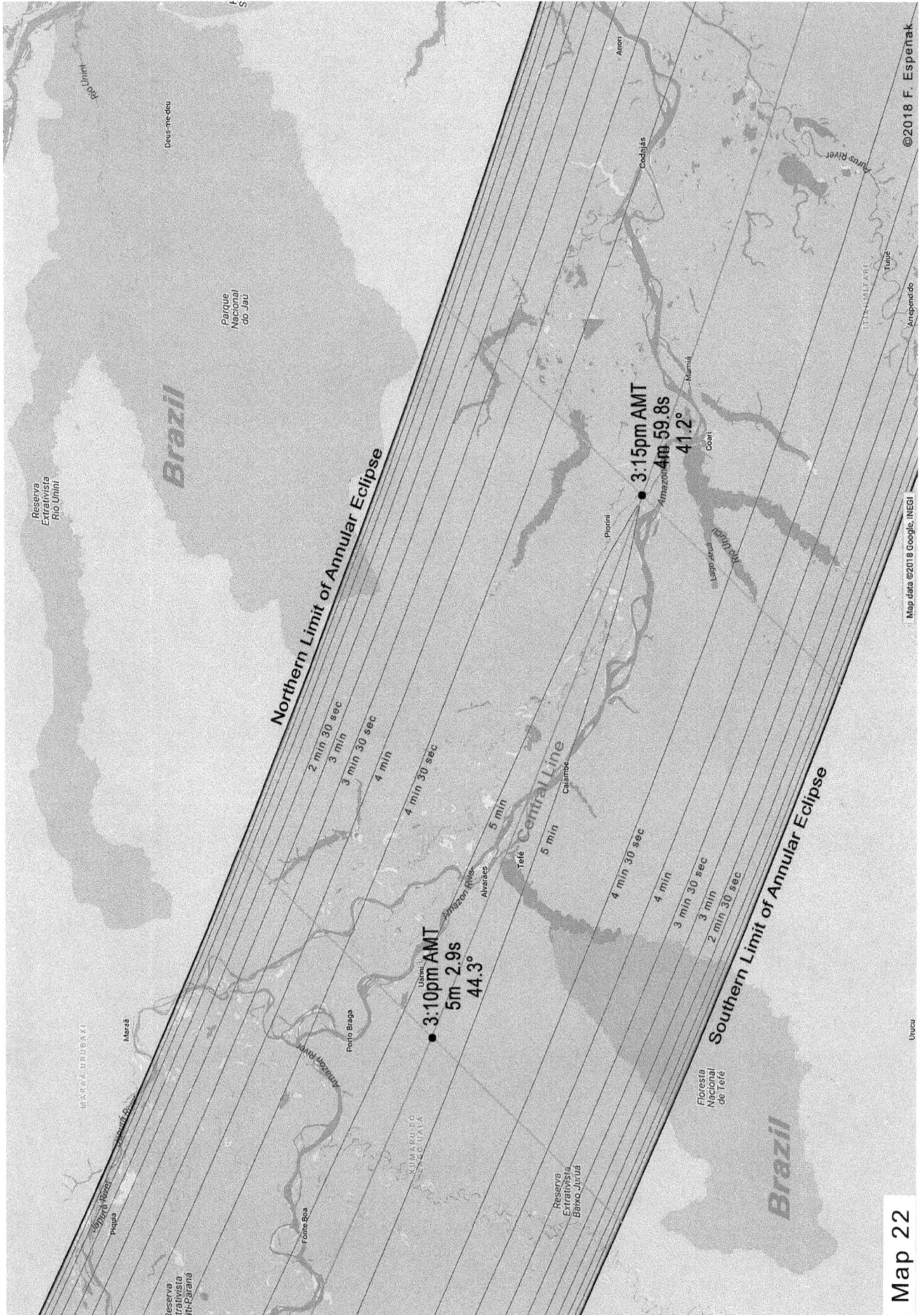

Northern Limit of Annular Eclipse

Southern Limit of Annular Eclipse

Central Line

Brazil

2 min 30 sec
3 min
3 min 30 sec
4 min
4 min 30 sec
5 min
5 min

5 min
4 min 30 sec
4 min
3 min 30 sec
3 min
2 min 30 sec

3:10pm AMT
5m 2.9s
44.3°

3:15pm AMT
4m 59.8s
41.2°

Parque Nacional do Jaú

Reserva Extrativista Rio Unini

Reserva Extrativista Baixo Juruá

Floresta Nacional de Tefé

MARAÃ URUBAXI

Amazon River

Amazon River

Purus River

Rio Uini

Deus-me-deu
Anori
Codajás
Manià
Coari
Piorini
Tefé
Alvarães
Caiambé
Uarini
Porto Braga
Fonte Boa
Marã
Tupé
Arapendido
Urucu

©2018 F. Espenak

Map data ©2018 Google, INEGI

Map 22

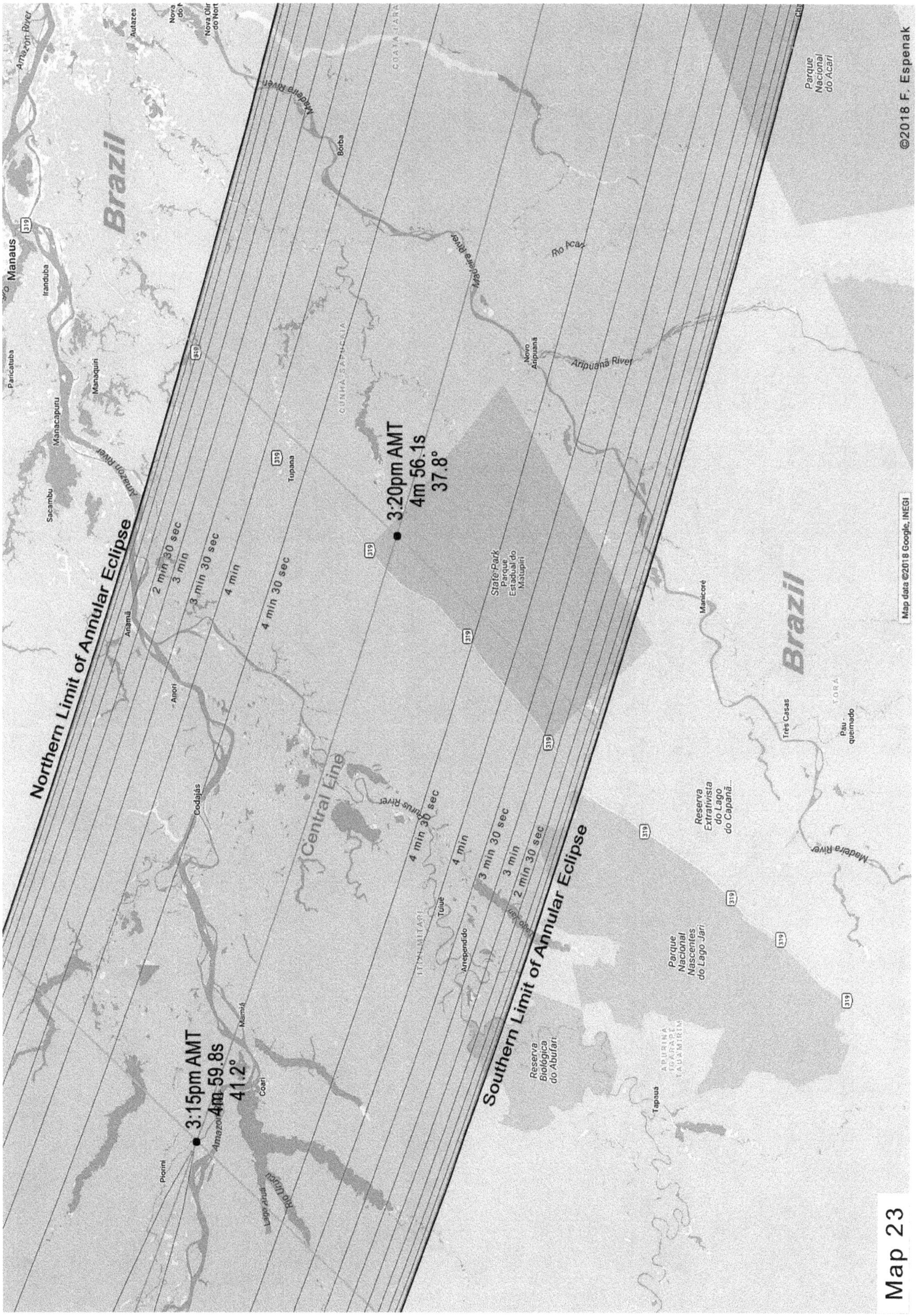

©2018 F. Espenak

Brazil

Parque Nacional
do Acari

Río Acari

Amazon River

Manaus

Iranduba

Panicatuba

Manacapuru

Sacambu

Manaquiri

Nova Olinda
do Norte

Autazes

Nazares

Borba

Madeira River

COATA-LARA

CUNHA-SAPUCAIA

Novo
Aripuanã

Aripuanã River

Northern Limit of Annular Eclipse

3:20pm AMT
4m 56.1s
37.8°

State-Park
Parque
Estadual do
Matupiri

2 min 30 sec
3 min
3 min 30 sec
4 min
4 min 30 sec

Aratana

Anori

Codajás

Central Line

Purus River

4 min 30 sec
4 min
3 min 30 sec
3 min
2 min 30 sec

ITAPIRANGA

Tupana

319

Maniçoré

Três Casas

Pau
queimado

YORA

Brazil

Reserva
Extrativista
do Lago
do Capanã

Madeira River

Tururí

Arependido

Southern Limit of Annular Eclipse

Parque
Nacional
Nascentes
do Lago Jari

319

319

319

3:15pm AMT
4m 59.8s
41.2°

Amazon River

Coari

Mamiá

Piorini

Lago Jauari

Río Puru

Reserva
Biológica
do Abufari

ARURINA
IGARAPÉ
TAUAMIRIM

Tapauá

Map data ©2018 Google, INEGI

Map 23

39

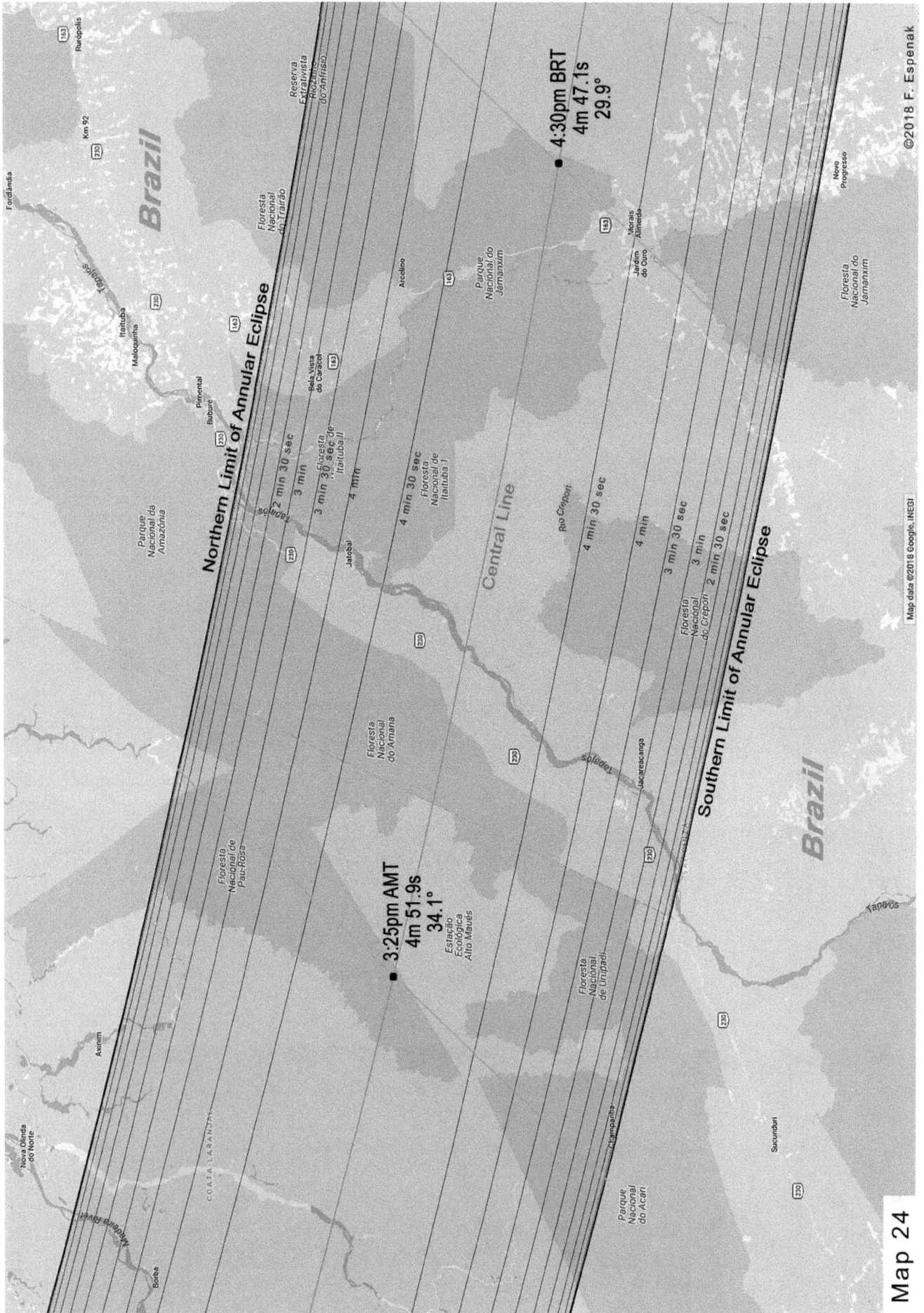

Map 24

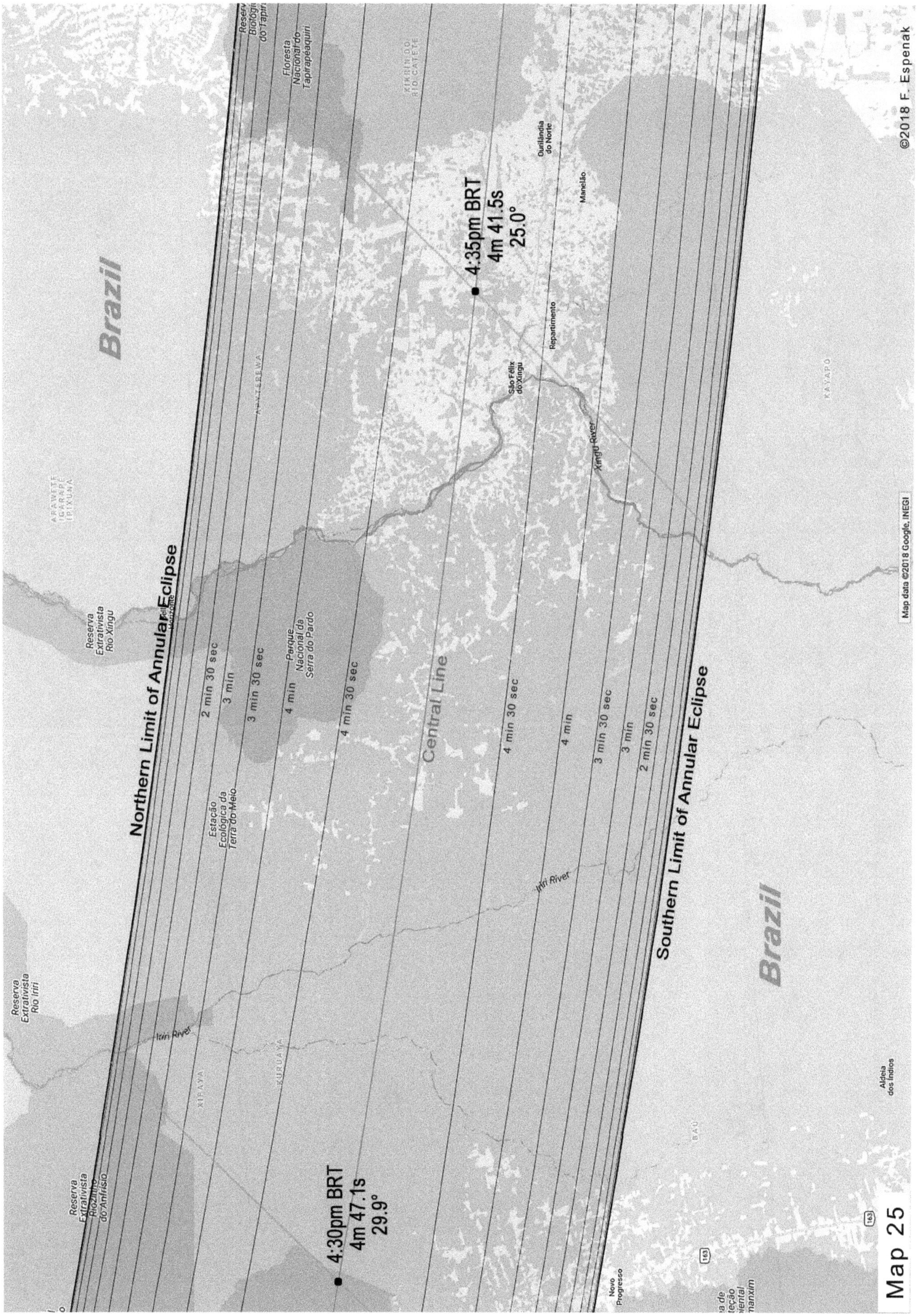
Map 25
©2018 F. Espenak
Map data ©2018 Google, INEGI
Brazil
Brazil
Northern Limit of Annular Eclipse
Southern Limit of Annular Eclipse
Central Line
4:30pm BRT
4m 47.1s
29.9°
4:35pm BRT
4m 41.5s
25.0°
2 min 30 sec
3 min
3 min 30 sec
4 min
4 min 30 sec
4 min 30 sec
4 min
3 min 30 sec
3 min
2 min 30 sec
Reserva Extrativista Rio Xingu
Reserva Extrativista Rio Iriri
Reserva Extrativista Riozinho do Anfrísio
Estação Ecológica da Terra do Meio
Parque Nacional da Serra do Pardo
Floresta Nacional do Tapirapé-aquiri
São Félix do Xingu
Repartimento
Ourilândia do Norte
Manelão
Novo Progresso
Aldeia dos Índios
Xingu River
Iriri River
Iriri River
KAYAPO
BAÚ
XIPAYA
KURUAYA
ARAWETE IGARAPÉ IPIXUNA
XIKRIN DO RIO CATETE
163
163

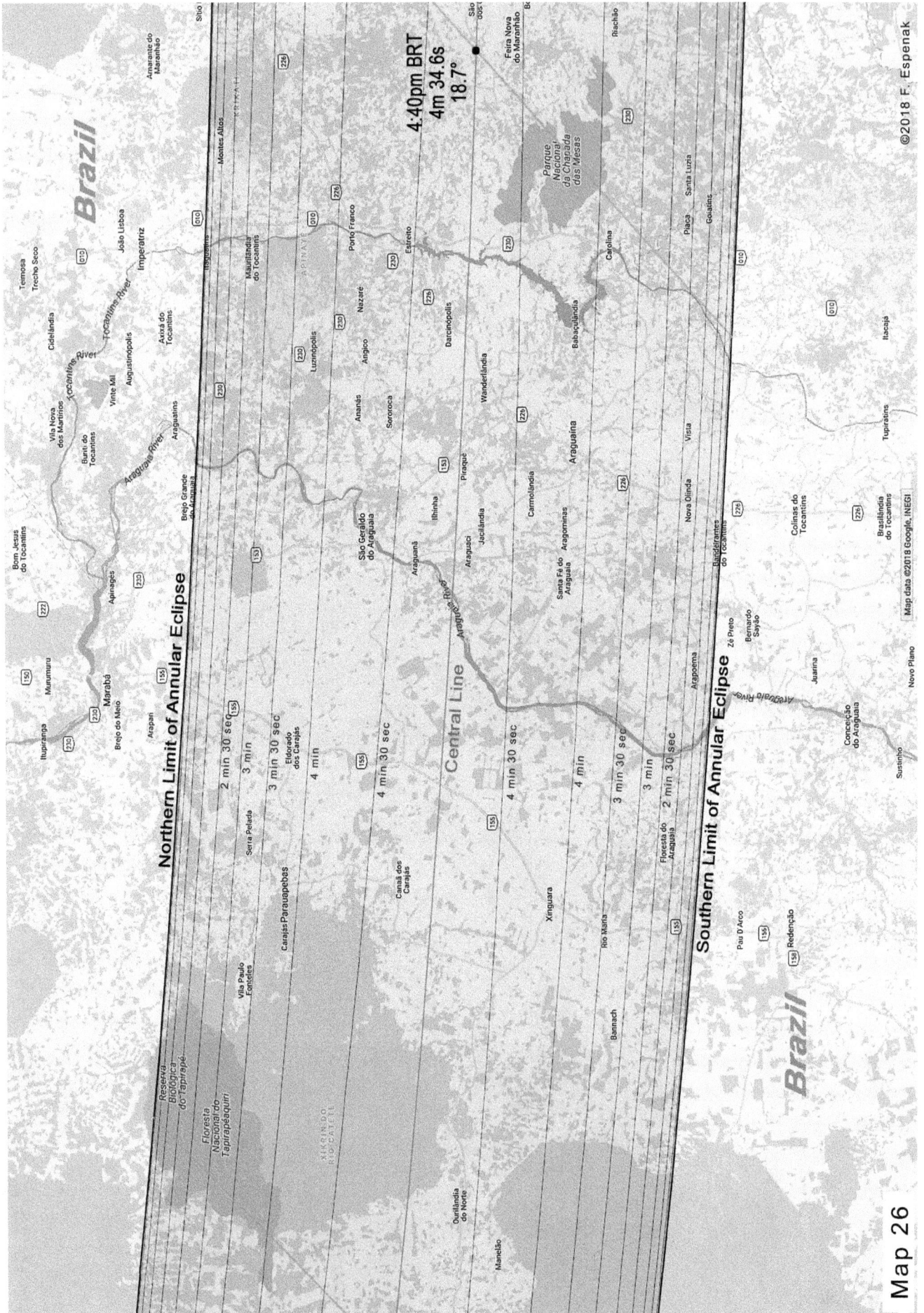

Map 26

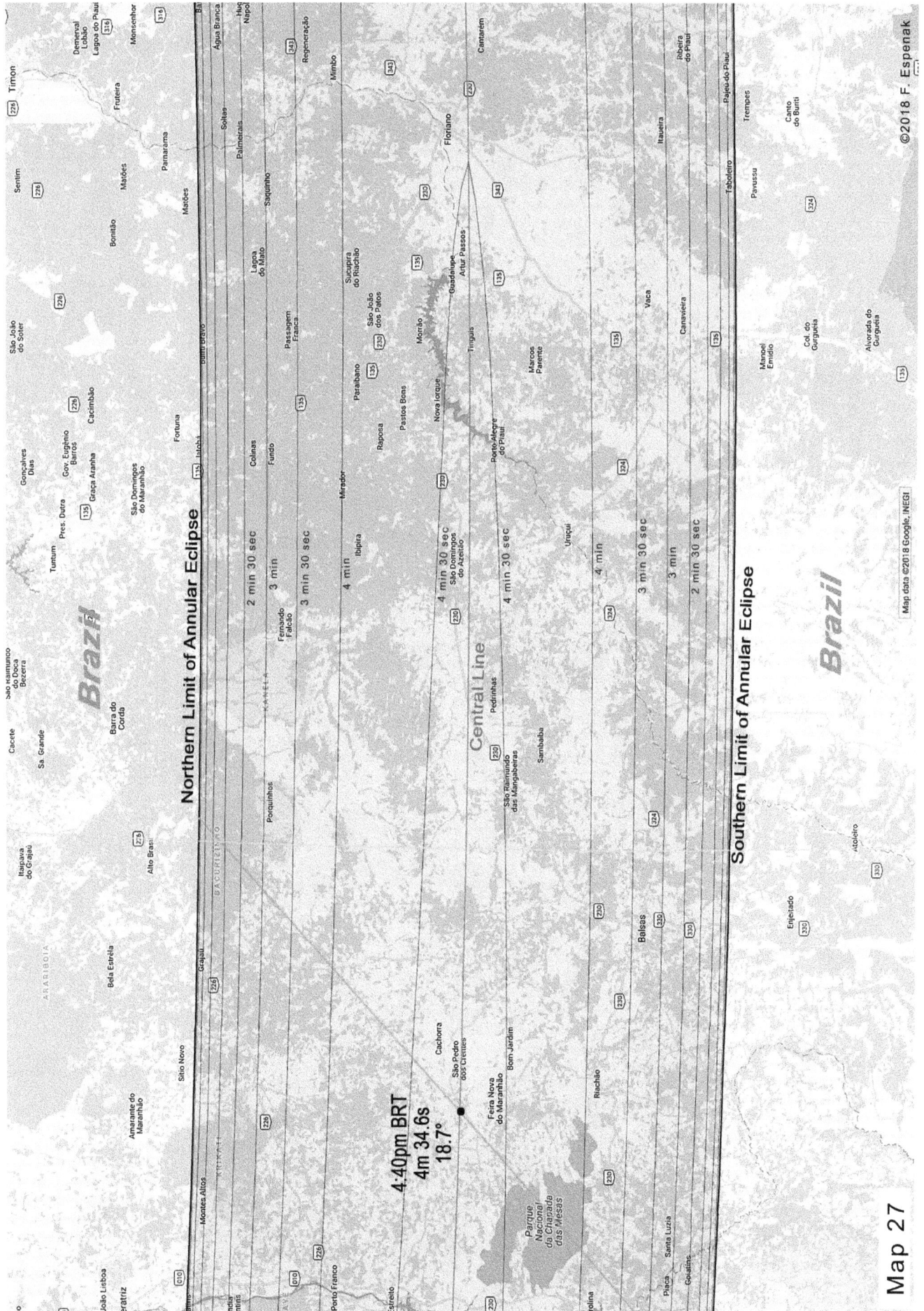

Northern Limit of Annular Eclipse

Central Line

Southern Limit of Annular Eclipse

Brazil

Brazil

4:40pm BRT
4m 34.6s
18.7°

2 min 30 sec
3 min
3 min 30 sec
4 min
4 min 30 sec

4 min 30 sec
4 min 30 sec
4 min
3 min 30 sec
3 min
2 min 30 sec

©2018 F. Espenak

Map data ©2018 Google, INEGI

Map 27

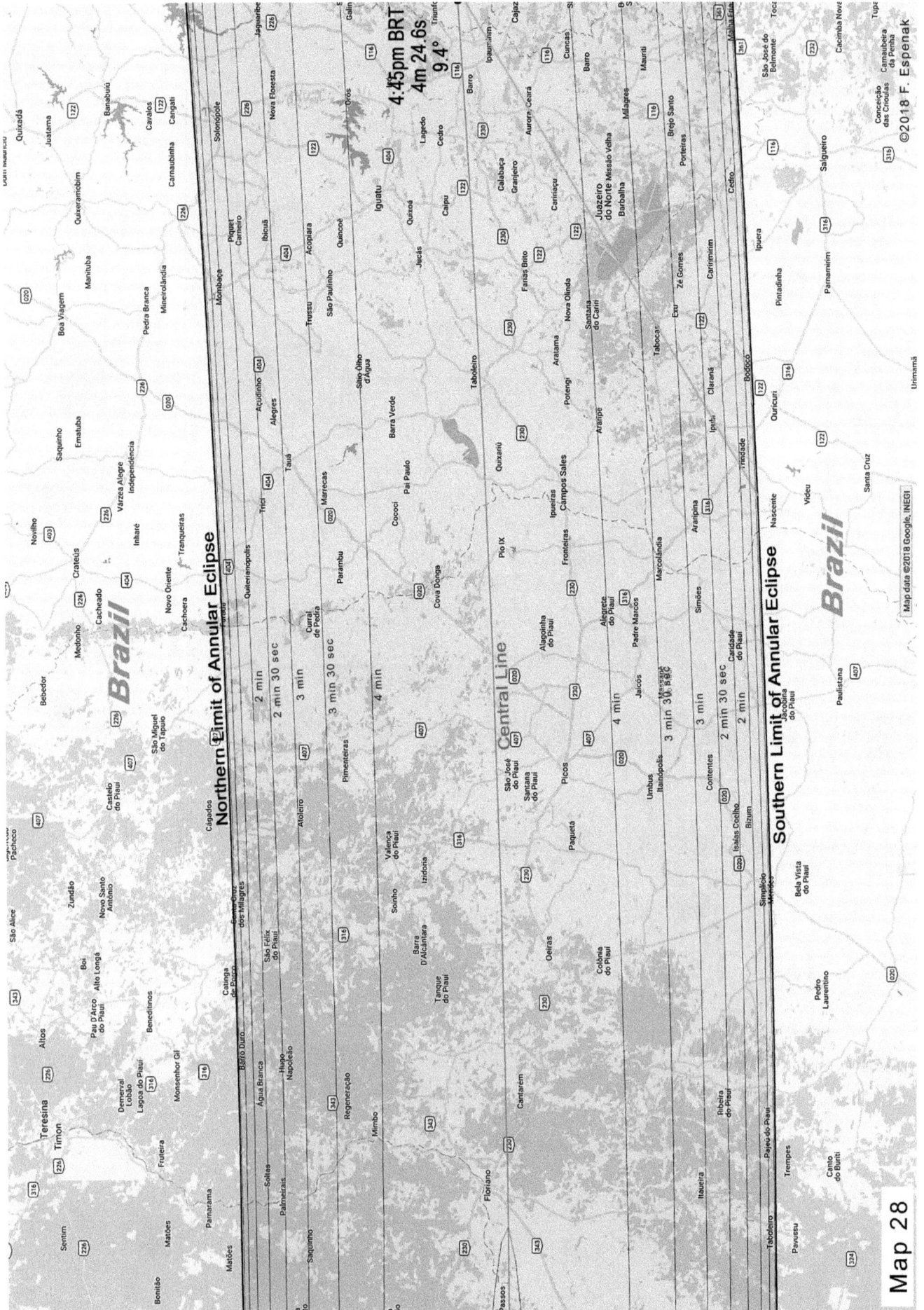

4:45pm BRT
4m 24.6s
9.4°

Brazil

Northern Limit of Annular Eclipse

Central Line

Southern Limit of Annular Eclipse

Brazil

2 min
2 min 30 sec
3 min
3 min 30 sec
4 min

4 min
3 min 30 sec
3 min
2 min 30 sec
2 min

©2018 F. Espenak

Map data ©2018 Google, INEGI

Map 28

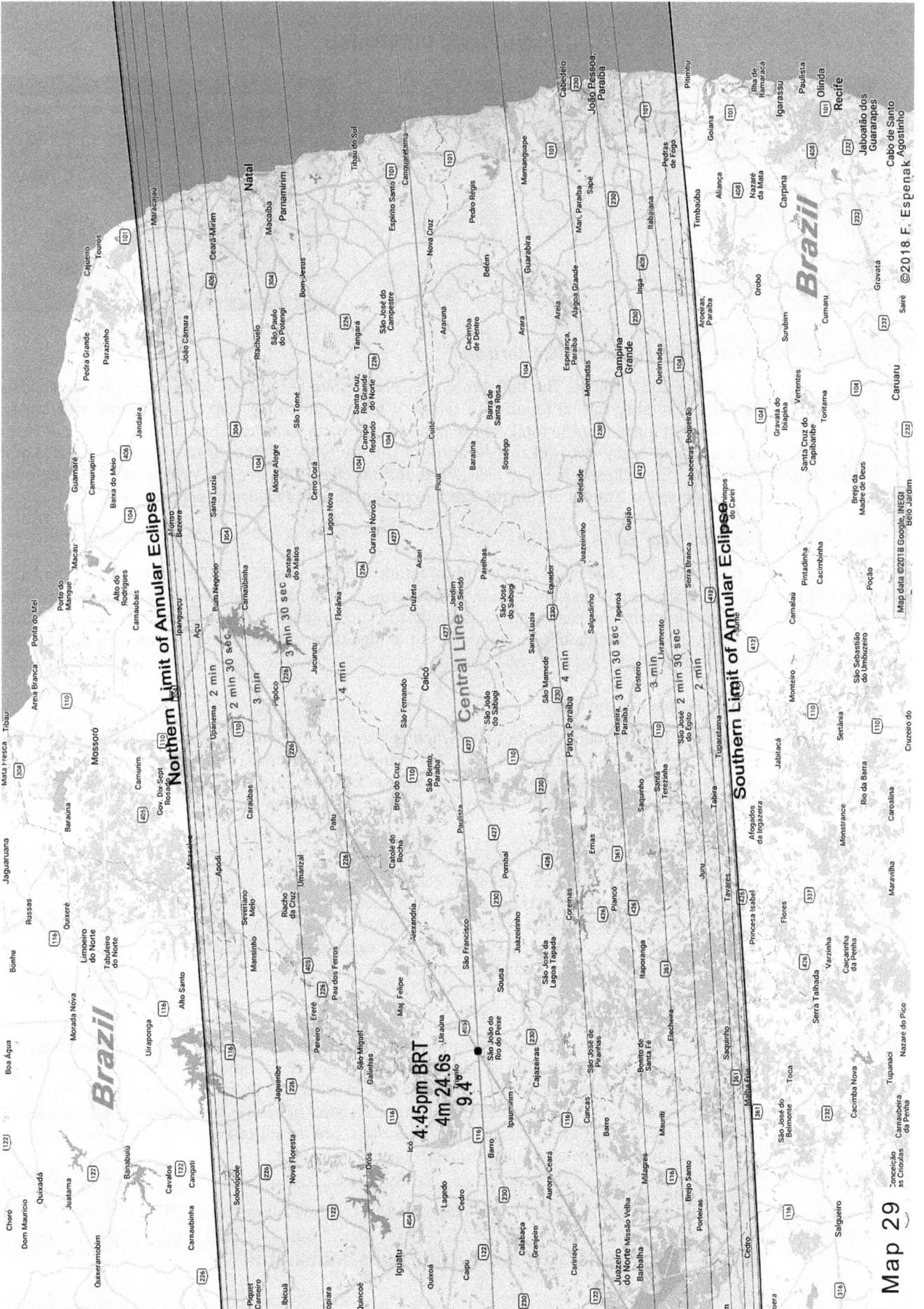

Brazil

Northern Limit of Annular Eclipse

Central Line

Southern Limit of Annular Eclipse

4:45pm BRT
4m 24.6s
9.4

Upanema 2 min
2 min 30 sec
3 min
3 min 30 sec
4 min
4 min
3 min 30 sec
3 min
2 min 30 sec
2 min

Brazil

©2018 F. Espenak

Map data ©2018 Google, INEGI

Map 29

45

AstroPixels Publishing

Eclipse Bulletin: Total Solar Eclipse of 2024 April 08

The ultimate guide to the 2024 eclipse. The bulletin is a treasure trove of facts on every aspect of the eclipse. Details about the path of totality can be found in a series of tables containing geographic coordinates, times, altitudes, and more. A set of maps plot the total eclipse path across North America. They show hundreds of cities and towns, major roads and highways, and the duration of totality with distance from the central line. Local circumstances tables give the eclipse times for nearly a thousand cites. The weather prospects along the path include cloud frequency maps and tables of climate statistics to help choose a favorable eclipse site.

Road Atlas for the Total Solar Eclipse of 2024

Contains 26 high resolution, full page maps of the path of totality across Mexico, the USA, and Canada. The large scale (1 inch = 22 miles) shows both major and minor roads, towns and cities, and rivers. Armed with this atlas and the latest weather forecasts, the road warrior is ready to chase totality no matter where it takes him/her along the 2500-mile-long path. This mobile strategy offers the highest probability of witnessing the spectacular 2024 total eclipse in clear skies.

Atlas of Central Solar Eclipses in the USA

When was the last total eclipse through the USA and when is the next? How often do they happen? What total eclipse tracks passed across the USA during the 17th, 18th, and 19th centuries, etc., and what states did they include? And how often is a total solar eclipse visible from each of the 50 states? The Atlas of Central Solar Eclipses in the USA answers all of these questions and more with hundreds of maps and tables.

21st Century Canon of Solar Eclipses

The complete guide with maps and data for all 224 solar eclipses occurring during the 100-year period from 2001 through 2100. A is comprehensive catalog lists the essential characteristics of each eclipse. A series of maps depict the geographic regions of visibility of each eclipse with 12 maps per page. There are full-page maps of every eclipse from 2017 through 2066. An appendix plots the track of every central eclipse (total, annular and hybrid) on large-scale maps with countries borders and major cities.

Thousand Year Canon of Solar Eclipses: 1501 – 2500

Contains maps and data for each of the 2,389 solar eclipses occurring over the ten century period centered on the present era. Some of the topics covered include eclipse classification, the visual appearance of each eclipse type, and eclipse predictions. The frequency each eclipse type, extremes in eclipse magnitude, greatest central duration are described. A comprehensive catalog lists the essential characteristics of each eclipse. An atlas of maps show the geographic regions of visibility of each eclipse. The 2,389 maps are arranged twelve to a page at an image scale permitting the assessment of eclipse visibility from any location on Earth.

All books are available it two editions: 1) Black and White, and 2) Color. For more information including sample pages of each, visit:

http://astropixels.com/pubs/index.html

www.ingramcontent.com/pod-product-compliance
Lightning Source LLC
Chambersburg PA
CBHW081423270326
41931CB00015B/3383